AQUATIC PLANTS: A PRACTICAL MANUAL

Dr. Kiran Choudhary

Oxford Book Company
Jaipur ▪ New Delhi

Published By

Oxford Book Company

Regd. Off. "Bony Residency", Gate No. 2
Opp. Tilak Public School, Vishweswariya Nagar,
Gopalpura Road, Jaipur - 302018, India
Ph: 0141 - 2761280, Telefax: 2761381
email - oxfordbookjpr@gmail.com
website - oxfordbookcompany.com

Delhi Office: 102, 1st Floor, Satyam House,
4327/3, Ansari Road, Darya Ganj,
New Delhi - 110002, India
Ph: 011-45652440

Distributed By

KEYA INTERNATIONAL PRESS LLP
E-mail: keyainternationalpressllp@gmail.com

First published in India, 2025

ISBN : 978-93-5524-559-5

Typeset by Shivangi Computers, Jaipur

Printed and bound in India by Thomson Press Ltd.,
New Delhi

Contents

Preface

Around the world, thousands of plant species can be found in freshwater habitats, such as the edges, surface, or bottom of shallow lakes and ponds; temporarily flooded low areas and meadows; seeps and springs (cienegas) in hill or montane regions; rooted in waterlogged soils; and along any other drainage system, whether man-made or natural. "Freshwater wetlands" are found below sea level and in certain very high alpine settings, where water may be either year-round or ephemeral. A freshwater wetland is often defined as one in which at least half of the species present are really aquatic plant species.

Numerous aquatic plant species are fundamentally cosmopolitan, which means that they are found all over the globe. Human activity is responsible for some of the most extensive dispersion. Although humans have unintentionally moved seeds, fruits, or vegetative clones from one pond or watershed to another, birds—especially waterfowl—are responsible for many of the cosmopolitan distributions. This is because the plant propagules are accidentally carried by the birds when they become stuck in their features or get stuck in mud on their feet. Early botanists could see many of the patterns shown by aquatic plants. Agnus Arber, for instance, documented many of the tactics we still discuss today in his 1920 book on aquatic plants. Low amounts of sunlight (PFD) are received by submerged leaves because light energy quickly decreases as it moves through a water column. In murky water with high algal surface populations, light penetration is very low. The segments of these underwater leaves, which are often so deeply dissected, may seemingly resemble macroscopic green algae (Chara and Nitella, for example). By having a correspondingly larger surface area, this technique maximizes surface-to-volume (S/V), allowing carbon dioxide to diffuse into the cells' chloroplasts more quickly. Because of their one- or two-cell layer structure, several aquatic organisms have very high surface to volume (S/V) ratios. Although the cuticle (wax) of these leaves is very thin, it is sufficiently porous to allow gases to diffuse through the surface with ease. Since water, not air, always envelops the photosynthetic organ of submerged plants, stomates—which are typically missing from these leaves—would be pointless. Given that the shoots are submerged in water, these leaves exhibit very low xylem tissue growth. Because of its higher specific gravity and poorly established intercellular air gaps, this plant is able to stay submerged. Water currents may pull and push on the finely dissected underwater shoot without causing any harm to the segments. These branches and leaves swing and dance wildly in fast-moving streams. In order to benefit from full light, floating leaves are often significantly wider, lack significant lobing, and stay flat on the water. For gas exchange, stomates are found, particularly on the top (adaxial) leaf surface. Because the cuticle on the top leaf surface is often quite noticeable, water may roll off without obstructing photosynthesis or encouraging the development of epiphytic algae. Chloroplasts may be abundant in the epidermis, and a bifacial mesophyll—palisade and spongy layers—is created. In addition to having hard cells called sclereids inside the mesophyll that give the leaf some toughness and keep the layers from collapsing, floating leaves often feature well-developed air chambers called lacunae that give them buoyancy. Numerous aquatic creatures that float are great subjects for research on logarithmic population increase. A single individual may be added to a pond and quickly grow vegetatively with full light and nonlimiting nutrients.

This book's goal is to provide readers with an understanding of the most prevalent aquatic plants that are found in and around water via descriptions and drawings.

— ***Dr. Kiran Choudhary***

Chapter 1

Introduction

Aquatic plants are plants that have adapted to living in aquatic environments (saltwater or freshwater). They are also referred to as hydrophytes or macrophytes. These plants require special adaptations for –living submerged in water, or at the water's surface - the most common adaptation is aerenchyma, but floating leaves and finely dissected leaves are also common. Aquatic plants can only grow in water or in soil that is permanently saturated with water. They are therefore a common component of wetlands.

The principal factor controlling the distribution of aquatic plants is the depth and duration of flooding. However, other factors may also control their distribution and abundance, including nutrients, disturbance from waves, grazing, and salinity. Aquatic vascular plants have originated on multiple occasions in different plant families; they can be ferns or angiosperms (including both monocots and dicots). Seaweeds are not vascular plants; rather they are multicellular marine algae, and therefore are not typically included among aquatic plants. A few aquatic plants are able to survive in brackish, saline, and salt water. Examples are found in genera such as Thalassia and Zostera. Although most aquatic plants can reproduce by flowering and setting seed, many also have extensive asexual reproduction by means of rhizomes, turions, and fragments in general.

One of the largest aquatic plants in the world is the Amazon water lily; one of the smallest is the minute duckweed. Many small aquatic animals use plants like duckweed for a home, or for protection from predators. But areas with more vegetation are likely to have more predators. Some other familiar examples of aquatic plants might include floating heart, water lily, lotus and water hyacinth.

Some aquatic plants are used by humans as a food source. Examples include wild rice (*Zizania*), water caltrop (*Trapa natans*), Chinese water chestnut (*Eleocharis dulcis*), Indian lotus (*Nelumbo nucifera*), water spinach (*Ipomoea aquatica*) and watercress (*Rorippa nasturtium-aquaticum*)

Morphological Classification

There are many possible classifications of aquatic plants based upon morphology. One example has six groups as follows:

- Amphiphytes: plants that are adapted to live either submerged or on land.
- Elodeids: stem plants that complete their entire life cycle submerged, or with only their flowers above the waterline.
- Isoetids: rosette plants that complete their entire life cycle submerged.
- Helophytes: plants rooted in the bottom, but with leaves above the waterline.
- Nymphaeids: plants rooted in the bottom, but with leaves floating on the water surface.
- Pleuston: vascular plants that float freely in the water.

Aquatic Ecosystem

An aquatic ecosystem is an ecosystem in a body of water. Communities of organisms that are dependent on each other and on their environment live in aquatic ecosystems. The two main types of aquatic ecosystems are marine ecosystems and freshwater ecosystems.

Types

Marine: Marine ecosystems cover approximately 71% of the Earth's surface and contain approximately 97% of the planet's water. They generate 32% of the world's net primary production. They are distinguished from freshwater ecosystems by the presence of dissolved compounds, especially salts, in the water. Approximately 85% of the dissolved materials in seawater are sodium and chlorine. Seawater has an average salinity of 35 parts per thousand (ppt) of water. Actual salinity varies among different marine ecosystems. Marine ecosystems can be divided into many zones depending upon water depth and shoreline features. The oceanic zone is the vast open part of the ocean where animals such as whales, sharks, and tuna live. The benthic zone consists of substrates below water where many invertebrates live. The intertidal zone is the area between high and low tides; in this figure it is termed the littoral zone. Other near-shore (neritic) zones can include estuaries, salt marshes, coral reefs, lagoons and mangrove swamps. In the deep water, hydrothermal vents may occur where chemosynthetic sulfur bacteria form the base of the food web. Classes of organisms found in marine ecosystems include brown algae, dinoflagellates, corals, cephalopods, echinoderms, and sharks. Fishes caught in marine ecosystems are the biggest source of commercial foods obtained from wild populations. Environmental problems concerning marine ecosystems include unsustainable exploitation of marine resources (for example overfishing of certain species), marine pollution, climate change, and building on coastal areas.

Freshwater

Freshwater ecosystems cover 0.80% of the Earth's surface and inhabit 0.009% of its total water. They generate nearly 3% of its net primary production. Freshwater ecosystems contain 41% of the world's known fish species.

There are three basic types of freshwater ecosystems:

- Lentic: slow moving water, including pools, ponds, and lakes.
- Lotic: faster moving water, for example streams and rivers.
- Wetlands: areas where the soil is saturated or inundated for at least part of the time.

Lentic

Lake ecosystems can be divided into zones. One common system divides lakes into three zones. The first, the littoral zone, is the shallow zone near the shore. This is where rooted wetland plants occur. The offshore is divided into two further zones, an open water zone and a deep water zone. In the open water zone (or photic zone) sunlight supports photosynthetic algae, and the species that feed upon them. In the deep water zone, sunlight is not available and the food web is based on detritus entering from the littoral and photic zones. Some systems use other names. The off shore areas may be called the pelagic zone, and the aphotic zone may be called the profundal zone. Inland from the littoral zone one can also frequently identify a riparian zone which has plants still affected by the presence of the lake—this can include effects from windfalls, spring flooding, and winter ice damage. The production of the lake as a whole is the result of production from plants growing in the littoral zone, combined with production from plankton growing in the open water. Wetlands can be part of the lentic system, as they form naturally along most lakeshores, the width of the wetland and littoral zone being dependent upon the slope of the shoreline and the amount of natural change in water levels, within and among years. Often dead trees accumulate in this zone, either from windfalls on the shore or logs transported to the site during floods. This woody debris provides important habitat for fish and nesting birds, as well as protecting shorelines from erosion, Two important subclasses of lakes are ponds, which typically are small lakes that intergrade with wetlands, and water reservoirs. Over long periods of time, lakes, or bays within them, may gradually become enriched by nutrients and slowly fill in with organic sediments, a process called succession. When humans use the watershed, the volumes of sediment entering the lake can accelerate this process. The addition of sediments and nutrients to a lake is known as eutrophication.

Lake Ecosystem

The ecosystem of a lake includes biotic (living) plants, animals and micro-organisms, as well as abiotic (nonliving) physical and chemical interactions. Lake ecosystems are a prime examples of lentic ecosystems. *Lentic* refers to standing or relatively still water, from the Latin *lentus*, which means sluggish. Lentic waters range from ponds to lakes to wetlands, and much of this article applies to lentic ecosystems in general. Lentic ecosystems can be compared with lotic ecosystems, which involve flowing terrestrial waters such as rivers and streams. Together, these two fields form the more general study area of freshwater or aquatic ecology.

Lentic systems are diverse, ranging from a small, temporary rainwater pool a few inches deep to Lake Baikal, which has a maximum depth of 1740 m. The general distinction between pools/ponds and lakes is vague, but Brown states that ponds and pools have their entire bottom surfaces exposed to light, while lakes do not. In addition, some lakes become seasonally stratified (discussed in more detail below.) Ponds and pools have two regions: the pelagic open water zone, and the benthic zone, which comprises the bottom and shore regions. Since lakes have deep bottom regions not exposed to light, these systems have an additional zone, the profundal. These three areas can have very different abiotic conditions and, hence, host species that are specifically adapted to live there.

Important Abiotic Factors

Light

Light provides the solar energy required to drive the process of photosynthesis, the major energy source of lentic systems (Brönmark and Hansson 2005). The amount of light received depends upon a combination of several factors. Small ponds may experience shading by surrounding trees, while cloud cover may affect light availability in all systems, regardless of size. Seasonal and diurnal considerations also play a role in light availability because the shallower the angle at which light strikes water, the more light is lost by reflection. This is known as Beers law (Giller and Malmqvist 1998).

Once light has penetrated the surface, it may also be scattered by particles suspended in the water column. This scattering decreases the total amount of light as depth increases (Moss 1998, Kalff 2002). Lakes are divided into photic and aphotic regions, the prior receiving sunlight and latter being below the depths of light penetration, making it void of photosynthetic capacity (Brönmark and Hansson 2005). In relation to lake zonation, the pelagic and benthic zones are considered to lie within the photic region, while the profundal zone is in the aphotic region (Brown 1987).

Temperature

Temperature is an important abiotic factor in lentic ecosystems because most of the biota are poikiloltheremic, where internal body temperatures are defined by the surrounding system. Water can be heated or cooled through radiation at the surface and conduction to or from the air and surrounding substrate (Giller and Malmqvist 1998). Shallow ponds often have a continuous temperature gradient from warmer waters at the surface to cooler waters at the bottom. In addition, temperature fluctuations can be very great in these systems, both diurnally and seasonally (Brown 1987).

Temperature regimes are very different in large lakes. In temperate regions, for example, as air temperatures increase, the icy layer formed on the surface of the lake breaks up, leaving the water at approximately 4 °C. This is the temperature at which water has the highest density. As the season progresses, the warmer air temperatures heat the surface waters, making them less dense. The deeper waters remain cool and dense due to reduced light penetration. As the summer begins, two distinct layers become established, with such a large temperature difference between them that they remain stratified. The lowest zone in the lake is the coldest and is called the hypolimnion. The upper warm zone is called the epilimnion. Between these zones is a band of rapid temperature change called the thermocline. During the colder fall season, heat is lost at the surface and the epilimnion cools. When the temperatures of the two zones are close enough, the waters begin to mix again to create a uniform temperature, an event termed lake turnover. In the winter, inverse stratification occurs as water near the surface cools freezes, while warmer, but denser water remains near the bottom. A thermocline is established, and the cycle repeats (Brown 1987, Brönmark and Hansson 2005).

Wind

In exposed systems, wind can create turbulent, spiral-formed surface currents called Langmuir circulations. Exactly how these currents become established is still not well understood, but it is evident that it involves some interaction between horizontal surface currents and surface gravity waves. The visible result of these rotations, which can be seen in any lake, are the surface foamlines that run parallel to the wind direction. Positively buoyant particles and small organisms concentrate in the foamline at the surface and negatively buoyant objects are found in the upwelling current between the two rotations. Objects with neutral buoyancy tend to be evenly distributed in the water column (Kalff 2002, Brönmark and Hansson 2005). This turbulence circulates nutrients in the water column, making it crucial for many pelagic species, however its effect on benthic and profundal organisms is minimal to non-existent, respectively (Kalff 2002). The degree of nutrient circulation is system specific, as it depends upon such factors as wind strength and duration, as well as lake or pool depth and productivity.

Chemistry

Oxygen is essential for organismal respiration. The amount of oxygen present in standing waters depends upon: 1) the area of transparent water exposed to the air, 2) the circulation of water within the system and 3) the amount of oxygen generated and used by organisms present (Brown 1987). In shallow, plant-rich pools there may be great fluctuations of oxygen, with extremely high concentrations occurring during the day due to photosynthesis and very low values at night when respiration is the dominant process of primary producers.

Thermal stratification in larger systems can also affect the amount of oxygen present in different zones. The epilimnion is oxygen rich because it circulates quickly, gaining oxygen via contact with the air. The hypolimnion, however, circulates very slowly and has no atmospheric contact. Additionally, fewer green plants exist in the hypolimnion, so there is less oxygen released from photosynthesis. In spring and fall when the epilimnion and hypolimnion mix, oxygen becomes more evenly distributed in the system (Brown 1987). Low oxygen levels are characteristic of the profundal zone due to the accumulation of decaying vegetation and animal matter that "rains" down from the pelagic and benthic zones and the inability to support primary producers (Brown 1987).

Phosphorus is important for all organisms because it is a component of DNA and RNA and is involved in cell metabolism as a component of ATP and ADP. Also, phosphorus is not found in large quantities in freshwater systems, limiting photosynthesis in primary producers, making it the main determinant of lentic system production. The phosphorus cycle is complex, but the model outlined below describes the basic pathways. Phosphorus mainly enters a pond or lake through runoff from the watershed or by atmospheric deposition. Upon entering the system, a reactive form of phosphorus is usually taken up by algae and macrophytes, which release a non-reactive phosphorus compound as a byproduct of photosynthesis. This phosphorus can drift downwards and become part of the benthic or profundal sediment, or it can be remineralized to the reactive form by microbes in the water column. Similarly, non-reactive phosphorus in the sediment can be remineralized into the reactive form (Brönmark and Hansson 2005). Sediments are generally richer in phosphorus than lake water, however, indicating that this nutrient may have a long residency time there before it is remineralized and re-introduced to the system (Kalff 2002).

Lentic System Biota

Bacteria

Bacteria are present in all regions of lentic waters. Free-living forms are associated with decomposing organic material, biofilm on the surfaces of rocks and plants, suspended in the water column, and in the sediments of the benthic and profundal zones. Other forms are also associated with the guts of lentic animals as parasites or in commensal relationships. Bacteria play an important role in system metabolism through nutrient recycling, which is discussed in the Trophic Relationships section.

Primary Producers

Algae, including both phytoplankton and periphyton are the principle photosynthesizers in ponds and lakes. Phytoplankton are found drifting in the water column of the pelagic zone. Many species have a higher density than water which should making them sink and end up in the benthos. To combat this, phytoplankton have developed density changing mechanisms, by forming vacuoles and gas vesicles or by changing their shapes to induce drag, slowing their descent. A very sophisticated adaptation utilized by a small number of species is a tail-like flagella that can adjust vertical position and allow movement in any direction. Phytoplankton can also maintain their presence in the water column by being circulated in Langmuir rotations. Periphytic algae, on the other hand, are attached to a substrate. In lakes and ponds, they can cover all benthic surfaces. Both types of plankton are important as food sources and as oxygen providers.

Aquatic plants live in both the benthic and pelagic zones and can be grouped according to their manner of growth: 1) emergent = rooted in the substrate but with leaves and flowers extending into the air, 2) floating-leaved = rooted in the substrate but with floating leaves, 3) submersed = growing beneath the surface and 4) free-floating macrophytes = not rooted in the substrate and floating on the surface. These various forms of macrophytes generally occur in different areas of the benthic zone, with emergent vegetation nearest the shoreline, then floating-leaved macrophytes, followed by submersed vegetation. Free-floating macrophytes can occur anywhere on the system's surface.

Aquatic plants are more buoyant than their terrestrial counterparts because freshwater has a higher density than air. This makes structural rigidity unimportant in lakes and ponds (except in the aerial stems and leaves).

Thus, the leaves and stems of most aquatic plants use less energy to construct and maintain woody tissue, investing that energy into fast growth instead. In order to contend with stresses induced by wind and waves, plants must be both flexible and tough. Light, water depth and substrate types are the most important factors controlling the distribution of submerged aquatic plants. Macrophytes are sources of food, oxygen, and habitat structure in the benthic zone, but cannot penetrate the depths of the euphotic zone and hence are not found there.

Invertebrates

Zooplankton are tiny animals suspended in the water column. Like phytoplankton, these species have developed mechanisms that keep them from sinking to deeper waters, including drag-inducing body forms and the active flicking of appendages such as antennae or spines. Remaining in the water column may have its advantages in terms of feeding, but this zone's lack of refugia leaves zooplankton vulnerable to predation. In response, some species, especially Daphnia sp., make daily vertical migrations in the water column by passively sinking to the darker lower depths during the day and actively moving towards the surface during the night. Also, because conditions in a lentic system can be quite variable across seasons, zooplankton have the ability to switch from laying regular eggs to resting eggs when there is a lack of food, temperatures fall below 2 °C, or if predator abundance is high. These resting eggs have a diapause, or dormancy period that should allow the zooplankton to encounter conditions that are more favourable to survival when they finally hatch.

The invertebrates that inhabit the benthic zone are numerically dominated by small species and are species rich compared to the zooplankton of the open water. They include Crustaceans (e.g. crabs, crayfish, and shrimp), molluscs (e.g. clams and snails), and numerous types of insects. These organisms are mostly found in the areas of macrophyte growth, where the richest resources, highly oxygenated water, and warmest portion of the ecosystem are found. The structurally diverse macrophyte beds are important sites for the accumulation of organic matter, and provide an ideal area for colonization. The sediments and plants also offer a great deal of protection from predatory fishes.

Very few invertebrates are able to inhabit the cold, dark, and oxygen poor profundal zone. Those that can are often red in color due to the presence of large amounts of hemoglobin, which greatly increases the amount of oxygen carried to cells. Because the concentration of oxygen within this zone is low, most species construct tunnels or borrows in which they can hide and make the minimum movements necessary to circulate water through, drawing oxygen to them without expending much energy.

Fishes and Other Vertebrates

Fishes have a range of physiological tolerances that are dependent upon which species they belong to. They have different lethal temperatures, dissolved oxygen requirements, and spawning needs that are based on their activity levels and behaviours. Because fishes are highly mobile, they are able to deal with unsuitable abiotic factors in one zone by simply moving to another. A detrital feeder in the profundal zone, for example, that finds the oxygen concentration has dropped too low may feed closer to the benthic zone. A fish might also alter its residence during different parts of its life history: hatching in a sediment nest, then moving to the weedy benthic zone to develop in a protected environment with food resources, and finally into the pelagic zone as an adult.

Other vertebrate taxa inhabit lentic systems as well. These include amphibians (e.g. salamanders and frogs), reptiles (e.g. snakes, turtles, and alligators), and a large number of waterfowl species.

Most of these vertebrates spend part of their time in terrestrial habitats and thus are not directly affected by abiotic factors in the lake or pond. Many fish species are important as consumers and as prey species to the larger vertebrates mentioned above.

Trophic Relationships

Primary Producers

Lentic systems gain most of their energy from photosynthesis performed by aquatic plants and algae. This autochthonous process involves the combination of carbon dioxide, water, and solar energy to produce carbohydrates and dissolved oxygen. Within a lake or pond, the potential rate of photosynthesis generally decreases with depth due to light attenuation. Photosynthesis, however, is often low at the top few millimeters of the surface, likely due to inhibition by ultraviolet light.

The exact depth and photosynthetic rate measurements of this curve are system specific and depend upon: 1) the total biomass of photosynthesizing cells, 2) the amount of light attenuating materials and 3) the abundance and frequency range of light absorbing pigments (i.e. chlorophylls) inside of photosynthesizing cells.

The energy created by these primary producers is important for the community because it is transferred to higher trophic levels via consumption.

Bacteria

The vast majority of bacteria in lakes and ponds obtain their energy by decomposing vegetation and animal matter. In the pelagic zone, dead fish and the occasional allochthonous input of litterfall are examples of coarse particulate organic matter (CPOM>1 mm). Bacteria degrade these into fine particulate organic matter (FPOM<1 mm) and then further into usable nutrients. Small organisms such as plankton are also characterized as FPOM.

Very low concentrations of nutrients are released during decomposition because the bacteria are utilizing them to build their own biomass. Bacteria, however, are consumed by protozoa, which are in turn consumed by zooplankton, and then further up the trophic levels. Nutrients, including those that contain carbon and phosphorus, are reintroduced into the water column at any number of points along this food chain via excretion or organism death, making them available again for bacteria. This regeneration cycle is known as the microbial loop and is a key component of lentic food webs.

The decomposition of organic materials can continue in the benthic and profundal zones if the matter falls through the water column before being completely digested by the pelagic bacteria. Bacteria are found in the greatest abundance here in sediments, where they are typically 2-1000 times more prevalent than in the water column.

Benthic Invertebrates

Benthic invertebrates, due to their high level of species richness, have many methods of prey capture. Filter feeders create currents via siphons or beating cilia, to pull water and its nutritional contents, towards themselves for straining. Grazers use scraping, rasping, and shredding adaptations to feed on periphytic algae and macrophytes. Members of the collector guild browse the sediments, picking out specific particles with raptorial appendages. Deposit feeding invertebrates indiscriminately consume sediment, digesting any organic material it contains. Finally, some invertebrates belong to the predator guild, capturing and consuming living animals. The profundal zone is home to a unique group of filter feeders that use small body movements to draw a current through burrows that they have created in the sediment. This mode of feeding requires the least amount of motion, allowing these species to conserve energy. A small number of invertebrate taxa are predators in the profundal zone. These species are likely from other regions and only come to these depths to feed. The vast majority of invertebrates in this zone are deposit feeders, getting their energy from the surrounding sediments.

Fish

Fish size, mobility, and sensory capabilities allow them to exploit a broad prey base, covering multiple zonation regions. Like invertebrates, fish feeding habits can be categorized into guilds. In the pelagic zone, herbivores graze on periphyton and macrophytes or pick phytoplankton out of the water column. Carnivores include fishes that feed on zooplankton in the water column (zooplanktivores), insects at the water's surface, on benthic structures, or in the sediment (insectivores), and those that feed on other fish (piscivores). Fish that consume detritus and gain energy by processing its organic material are called detritivores. Omnivores ingest a wide variety of prey, encompassing floral, faunal, and detrital material. Finally, members of the parasitic guild acquire nutrition from a host species, usually another fish or large vertebrate. Fish taxa are flexible in their feeding roles, varying their diets with environmental conditions and prey availability. Many species also undergo a diet shift as they develop. Therefore, it is likely that any single fish occupies multiple feeding guilds within its lifetime.

Lentic Food Webs

As noted in the previous sections, the lentic biota are linked in complex web of trophic relationships. These organisms can be considered to loosely be associated with specific trophic groups (e.g. primary producers, herbivores, primary carnivores, secondary carnivores, etc.). Scientists have developed several theories in order to understand the mechanisms that control the abundance and diversity within these groups. Very generally, top-down processes dictate that the abundance of prey taxa is dependent upon the actions of consumers from higher trophic levels. Typically, these processes operate only between two trophic levels, with no effect on the others. In some cases, however, aquatic systems experience a trophic cascade; for example, this might occur if primary producers experience less grazing by herbivores because these herbivores are suppressed by carnivores.

Bottom-up processes are functioning when the abundance or diversity of members of higher trophic levels is dependent upon the availability or quality of resources from lower levels. Finally, a combined regulating theory, bottom-up:top-down, combines the predicted influences of consumers and resource availability. It predicts that trophic levels close to the lowest trophic levels will be most influenced by bottom-up forces, while top-down effects should be strongest at top levels.

Community patterns and diversity

Local Species Richness

The biodiversity of a lentic system increases with the surface area of the lake or pond. This is attributable to the higher likelihood of partly terrestrial species of finding a larger system. Also, because larger systems typically have larger populations, the chance of extinction is decreased.

Additional factors, including temperature regime, pH, nutrient availability, habitat complexity, speciation rates, competition, and predation, have been linked to the number of species present within systems.

Succession Patterns in Plankton Communities – the PEG Model

Phytoplankton and zooplankton communities in lake systems undergo seasonal succession in relation to nutrient availability, predation, and competition. Sommer *et al.* described these patterns as part of the Plankton Ecology Group (PEG) model, with 24 statements constructed from the analysis of numerous systems. The following includes a subset of these statements, as explained by Brönmark and Hansson illustrating succession through a single seasonal cycle:

Winter

- Increased nutrient and light availability result in rapid phytoplankton growth towards the end of winter. The dominant species, such as diatoms, are small and have quick growth capabilities.

- These plankton are consumed by zooplankton, which become the dominant plankton taxa.

Spring

- A clear water phase occurs, as phytoplankton populations become depleted due to increased predation by growing numbers of zooplankton.

Summer

- Zooplankton abundance declines as a result of decreased phytoplankton prey and increased predation by juvenile fishes.
- With increased nutrient availability and decreased predation from zooplankton, a diverse phytoplankton community develops.
- As the summer continues, nutrients become depleted in a predictable order: phosphorus, silica, and then nitrogen. The abundance of various phytoplankton species varies in relation to their biological need for these nutrients.
- Small-sized zooplankton become the dominant type of zooplankton because they are less vulnerable to fish predation.

Fall

- Predation by fishes is reduced due to lower temperatures and zooplankton of all sizes increase in number.

Winter

- Cold temperatures and decreased light availability result in lower rates of primary production and decreased phytoplankton populations. Reproduction in zooplankton decreases due to lower temperatures and less prey.

The PEG model presents an idealized version of this succession pattern, while natural systems are known for their variation.

Latitudinal Patterns

There is a well-documented global pattern that correlates decreasing plant and animal diversity with increasing latitude, that is to say, there are fewer species as one moves towards the poles. The cause of this pattern is one of the greatest puzzles for ecologists today. Theories for its explanation include energy availability, climatic variability, disturbance, competition, etc. Despite this global diversity gradient, this pattern can be weak for freshwater systems compared to global marine and terrestrial systems. This may be related to size, as Hillebrand and Azovsky found that smaller organisms (protozoa and plankton) did not follow the expected trend strongly, while larger species (vertebrates) did. They attributed this to better dispersal ability by smaller organisms, which may result in high distributions globally.

Natural Lake Lifecycles

Lake Creation

Lakes can be formed in a variety of ways, but the most common are discussed briefly below. The oldest and largest systems are the result of tectonic activities. The rift lakes in Africa, for example are the result of seismic activity along the site of separation of two tectonic plates. Ice-formed lakes are created when glaciers recede, leaving behind abnormalities in the landscape shape that are then filled with water. Finally, oxbow lakes are fluvial in origin, resulting when a meandering river bend is pinched off from the main channel.

Natural Extinction

All lakes and ponds receive sediment inputs. Since these systems are not really expanding, it is logical to assume that they will become increasingly shallower in depth, eventually becoming wetlands or terrestrial vegetation. The length of this process should depend upon a combination of depth and sedimentation rate. Moss gives the example of Lake Tanganyika, which reaches a depth of 1500 m and has a sedimentation rate of 0.5 mm/yr. Assuming that sedimentation is not influenced by anthropogenic factors, this system should go extinct in approximately 3 million years. Shallow lentic systems might also fill in as swamps encroach inward from the edges. These processes operate on a much shorter timescale, taking hundreds to thousands of years to complete the extinction process.

Human Impacts

Acidification

Sulfur dioxide and nitrogen oxides are naturally released from volcanoes, organic compounds in the soil, wetlands, and marine systems, but the majority of these compounds come from the combustion of coal, oil, gasoline, and the smelting of ores containing sulfur. These substances dissolve in atmospheric moisture and enter lentic systems as acid rain. Lakes and ponds that contain bedrock that is rich in carbonates have a natural buffer, resulting in no alteration of pH. Systems without this bedrock, however, are very sensitive to acid inputs because they have a low neutralizing capacity, resulting in pH declines even with only small inputs of acid. At a pH of 5–6 algal species diversity and biomass decrease considerably, leading to an increase in water transparency – a characteristic feature of acidified lakes. As the pH continues lower, all fauna becomes less diverse. The most significant feature is the disruption of fish reproduction. Thus, the population is eventually composed of few, old individuals that eventually die and leave the systems without fishes. Acid rain has been especially harmful to lakes in Scandinavia, western Scotland, west Wales and the north eastern United States.

Eutrophication

Eutrophic systems contain a high concentration of phosphorus (~30+μg/L), nitrogen (~1500+μg/L), or both. Phosphorus enters lentic waters from wastewater treatment effluents, discharge from raw sewage, or from runoff of farmland. Nitrogen mostly comes from agricultural fertilizers from runoff or leaching and subsequent groundwater flow.

This increase in nutrients required for primary producers results in a massive increase of phytoplankton growth, termed a plankton bloom. This bloom decreases water transparency, leading to the loss of submerged plants. The resultant reduction in habitat structure has negative impacts on the species' that utilize it for spawning, maturation and general survival. Additionally, the large number of short-lived phytoplankton result in a massive amount of dead biomass settling into the sediment. Bacteria need large amounts of oxygen to decompose this material, reducing the oxygen concentration of the water. This is especially pronounced in stratified lakes when the thermocline prevents oxygen rich water from the surface to mix with lower levels. Low or anoxic conditions preclude the existence of many taxa that are not physiologically tolerant of these conditions.

Invasive Species

Invasive species have been introduced to lentic systems through both purposeful events (e.g. stocking game and food species) as well as unintentional events (e.g. in ballast water). These organisms can affect natives via competition for prey or habitat, predation, habitat alteration, hybridization, or the introduction of harmful diseases and parasites.

With regard to native species, invaders may cause changes in size and age structure, distribution, density, population growth, and may even drive populations to extinction. Examples of prominent invaders of lentic systems include the zebra mussel and sea lamprey in the Great Lakes.

Ponds

Ponds are small pools with shallow water, marsh, and aquatic plants. They can be further divided into four zones: vegetation zone, open water, bottom mud and surface film. The size and depth of ponds often varies greatly with the time of year; many ponds are produced by spring flooding from rivers. Food webs are based both on free-floating algae and upon aquatic plants. There is usually a diverse array of aquatic life, with a few examples including algae, snails, fish, beetles, water bugs, frogs, turtles, otters and muskrats. Top predators may include large fish, herons, or alligators.

Since fish are a major predator upon amphibian larvae, ponds that dry up each year, thereby killing resident fish, provide important refugia for amphibian breeding. Ponds that dry up completely each year are often known as vernal pools. Some ponds are produced by animal activity, including alligator holes and beaver ponds, and these add important diversity to landscapes.

Lotic

The major zones in river ecosystems are determined by the river bed's gradient or by the velocity of the current. Faster moving turbulent water typically contains greater concentrations of dissolved oxygen, which supports greater biodiversity than the slow moving water of pools. These distinctions forms the basis for the division of rivers into upland and lowland rivers.

The food base of streams within riparian forests is mostly derived from the trees, but wider streams and those that lack a canopy derive the majority of their food base from algae. Anadromous fish are also an important source of nutrients. Environmental threats to rivers include loss of water, dams, chemical pollution and introduced species. A dam produces negative effects that continue down the watershed. The most important negative effects are the reduction of spring flooding, which damages wetlands, and the retention of sediment, which leads to loss of deltaic wetlands.

River Ecosystem

The ecosystem of a river is the river viewed as a system operating in its natural environment, and includes biotic (living) interactions amongst plants, animals and micro-organisms, as well as abiotic (nonliving) physical and chemical interactions.

River ecosystems are prime examples of lotic ecosystems. *Lotic* refers to flowing water, from the Latin *lotus*, washed. Lotic waters range from springs only a few centimeters wide to major rivers kilometers in width. Much of this article applies to lotic ecosystems in general, including related lotic systems such as streams and springs. Lotic ecosystems can be contrasted with lentic ecosystems, which involve relatively still terrestrial waters such as lakes and ponds. Together, these two fields form the more general study area of freshwater or aquatic ecology.

The following unifying characteristics make the ecology of running waters unique from that of other aquatic habitats.

- Flow is unidirectional.
- There is a state of continuous physical change.
- There is a high degree of spatial and temporal heterogeneity at all scales (microhabitats).

- Variability between lotic systems is quite high.
- The biota is specialized to live with flow conditions.

Wetlands

Wetlands are dominated by vascular plants that have adapted to saturated soil. There are four main types of wetlands: swamp, marsh, fen and bog (both fens and bogs are types of mire). Wetlands are the most productive natural ecosystems in the world because of the proximity of water and soil. Hence they support large numbers of plant and animal species. Due to their productivity, wetlands are often converted into dry land with dykes and drains and used for agricultural purposes. The construction of dykes, and dams, has negative consequences for individual wetlands and entire watersheds. Their closeness to lakes and rivers means that they are often developed for human settlement. Once settlements are constructed and protected by dykes, the settlements then become vulnerable to land subsidence and ever increasing risk of flooding. The Louisiana coast around New Orleans is a well-known example; the Danube Delta in Europe is another.

Functions

Aquatic ecosystems perform many important environmental functions. For example, they recycle nutrients, purify water, attenuate floods, recharge ground water and provide habitats for wildlife. Aquatic ecosystems are also used for human recreation, and are very important to the tourism industry, especially in coastal regions.

The health of an aquatic ecosystem is degraded when the ecosystem's ability to absorb a stress has been exceeded. A stress on an aquatic ecosystem can be a result of physical, chemical or biological alterations of the environment. Physical alterations include changes in water temperature, water flow and light availability.

Chemical alterations include changes in the loading rates of biostimulatory nutrients, oxygen consuming materials, and toxins. Biological alterations include over-harvesting of commercial species and the introduction of exotic species. Human populations can impose excessive stresses on aquatic ecosystems. There are many examples of excessive stresses with negative consequences. Consider three. The environmental history of the Great Lakes of North America illustrates this problem, particularly how multiple stresses, such as water pollution, over-harvesting and invasive species can combine. The Norfolk Broadlands in England illustrate similar decline with pollution and invasive species. Lake Pontchartrain along the Gulf of Mexico illustrates the negative effects of different stresses including levee construction, logging of swamps, invasive species and salt water intrusion.

Abiotic Characteristics

An ecosystem is composed of biotic communities that are structured by biological interactions and abiotic environmental factors. Some of the important abiotic environmental factors of aquatic ecosystems include substrate type, water depth, nutrient levels, temperature, salinity, and flow. It is often difficult to determine the relative importance of these factors without rather large experiments. There may be complicated feed back loops. For example, sediment may determine the presence of aquatic plants, but aquatic plants may also trap sediment, and add to the sediment through peat.

The amount of dissolved oxygen in a water body is frequently the key substance in determining the extent and kinds of organic life in the water body. Fish need dissolved oxygen to survive, although their tolerance to low oxygen varies among species; in extreme cases of low oxygen some fish even resort to air gulping. Plants often have to produce aerenchyma, while the shape and size of leaves may also be altered. Conversely, oxygen is fatal to many kinds of anaerobic bacteria.

Nutrient levels are important in controlling the abundance of many species of algae. The relative abundance of nitrogen and phosphorus can affect determine which species of algae come to dominate. Algae are a very important source of food for aquatic life, but at the same time, if they become over-abundant, they can cause declines in fish when they decay. Similar over-abundance of algae in coastal environments such as the Gulf of Mexico produces, upon decay, a hypoxic region of water known as a dead zone.

The salinity of the water body is also a determining factor in the kinds of species found in the water body. Organisms in marine ecosystems tolerate salinity, while many freshwater organisms are intolerant of salt. The degree of salinity in an estuary or delta may is an important control upon the type of wetland (fresh, intermediate, or brackish), and the associated animal species. Dams built upstream may reduce spring flooding, and reduce sediment accretion, and may therefore lead to saltwater intrusion in coastal wetlands. Freshwater used for irrigation purposes often absorb levels of salt that are harmful to freshwater organisms.

Biotic Characteristics

The biotic characteristics are mainly determined by the organisms that occur. For example, wetland plants may produce dense canopies that cover large areas of sediment—or snails or geese may graze the vegetation leaving large mud flats. Aquatic environments have relatively low oxygen levels, forcing adaptation by the organisms found there. For example, many wetland plants must produce aerenchyma to carry oxygen to roots. Other biotic characteristics are more subtle and difficult to measure, such as the relative importance of competition, mutualism or predation. There are a growing number of cases where predation by coastal herbivores including snails, geese and mammals appears to be a dominant biotic factor.

Autotrophic Organisms

Autotrophic organisms are producers that generate organic compounds from inorganic material. Algae use solar energy to generate biomass from carbon dioxide and are possibly the most important autotrophic organisms in aquatic environments. Of course, the more shallow the water, the greater the biomass contribution from rooted and floating vascular plants.

These two sources combine to produce the extraordinary production of estuaries and wetlands, as this autotrophic biomass is converted into fish, birds, amphibians and other aquatic species.

Chemosynthetic bacteria are found in benthic marine ecosystems. These organisms are able to feed on hydrogen sulfide in water that comes from volcanic vents. Great concentrations of animals that feed on this bacteria are found around volcanic vents. For example, there are giant tube worms

(*Riftia pachyptila*) 1.5m in length and clams (*Calyptogena magnifica*) 30 cm long.

Heterotrophic Organisms

Heterotrophic organisms consume autotrophic organisms and use the organic compounds in their bodies as energy sources and as raw materials to create their own biomass. Euryhaline organisms are salt tolerant and can survive in marine ecosystems, while stenohaline or salt intolerant species can only live in freshwater environments.

Plant Identification

This section of AQUAPLANT is designed to help you identify aquatic plants using photos, drawings, and descriptions. Using this section involves viewing photos and drawings of aquatic plants until you identify the plant in question.

Aquatic plants are generally divided into four groups for management purposes. These groups are the Algae, Floating Plants, Submerged Plants, and Emergent Plants.

Algae are very primitive plants. Some algae are microscopic (planktonic algae), others are thin and stringy or hair-like (filamentous algae), while still others are large and resemble higher plants but without true roots (chara).

True Floating Plants are not attached to the bottom. Floating plants come in sizes from very small (duckweed) to over a foot in diameter (water hyacinth). Most have roots that hang in the water from the floating green portions.

Submerged Plants are rooted plants with most of their vegetative mass below the water surface, although some portions may stick above the water. One discerning characteristic of submerged plants is their flaccid or soft stems, which is why they do not usually rise above the water's surface.

Emergent Plants are rooted plants often along the shoreline that stand above the surface of the water (cattails). The stems of emergent plants are somewhat stiff or firm.

Many ponds have more than one type of aquatic plant, and care must be taken to identify all the aquatic plants inhabiting the pond. Some pond plants may be beneficial to local or migratory wildlife, and therefore, may want to be encouraged or at least not eliminated. Click on whichever group of aquatic plants that you feel your specimen may belong to and work through the examples until you find it.

- Amphiphytes: plants that are adapted to live either submerged or on land.
- Elodeids: stem plants that complete their entire life cycle submerged, or with only their flowers above the waterline.
- Isoetids: rosette plants that complete their entire life cycle submerged.
- Helophytes: plants rooted in the bottom, but with leaves above the waterline.
- Nymphaeids: plants rooted in the bottom, but with leaves floating on the water surface.
- Pleuston: vascular plants that float freely in the water.

Chapter 2

The Need of Aquatic Plants

Aquatic plants are an often misunderstood and undervalued part of freshwater ecosystems. In fact, many people would rather not have them in their favourite swimming or fishing hole. The following points illustrate a few of the important roles that aquatic plants play in a water body.

Food

Aquatic plants provide important food for many animals. Ducks and geese eat the seeds, leafy parts, and tubers of plants such as pondweeds (*Potomogeten* spp.), watershield (*Brasenia schreberi*), arrowhead (*Sagittaria latifolia)*, water pepper (*Polygonum* sp.), and duckweed (*Lemna* sp.). Songbirds use fluff from cattails (*Typha* sp.) as nest material and eat the seeds of many emergent plants. Otter, beaver, muskrats, turtles, and moose will also graze on a variety of aquatic plants. More food for fish is produced in areas of aquatic vegetation than in areas where there are no plants. Insect larvae, snails, and freshwater shrimp thrive in plant beds. Sunfish—Minnesota's most sought-after game fish—eat aquatic plants in addition to aquatic insects and crustaceans.

Historically humans have also utilized aquatic plants as a food source. Cattails have edible shoots and roots and even the pollen has been used in making biscuits. Arrowheads form large edible tubers at the root ends, called duck potatoes, which were consumed by Native Americans. Watercress (*Rorippa nasturtium-aquaticum*) has many historic medicinal uses and its spicy vegetation continues to be used in salads and garnishes. Water lily roots are a common source of food in many parts of the world and have historic medicinal value. Even the submersed plant, coontail (*Ceratophylum demersum*) has been used for medicinal purposes.

Habitat

Aquatic plants provide important living space for small animals such as aquatic insects, snails, and crustaceans, which in turn supply food for fish and waterfowl. Many studies have shown that vegetated areas support many times more of these tiny creatures than do unvegetated areas.

Plants provide shelter for young fish. Because bass, sunfish, and yellow perch usually nest in areas where vegetation is growing, certain areas of lakes are protected and posted by the DNR as fish spawning areas during spring and early summer. Northern pike use aquatic plants, too, by spawning in marshy and flooded areas in early spring.

Cover

Young fish and amphibians will use aquatic plants as a source of cover from predatory fish and birds. This, coupled with the abundant food supply, makes aquatic plants important nurseries for baby fish (including our native salmon), frogs, and salamanders.

Housing Supplies

The sturdy emergent plants provide nest and den-building materials for many birds and mammals, including muskrats.

Humans also construct baskets, mats, boats, and even dwellings from cattail, rush, and bulrush stems.

Erosion Control

Submersed and emergent plants protect shorelines from erosion due to wave action or currents. They can also help stabilize the sediment which can increase water clarity.

Nutrient Cycling

Aquatic plants form a vital part of the complex system of chemical cycling in a waterbody. They can also influence the supply of oxygen in the water. Recently aquatic plants have received a lot of attention for their ability to soak up pollutants from contaminated water. They utilize nutrients that would otherwise be used by algae, thereby improving water clarity. Increasing attention is being paid toward their possible use as indicators of water quality.

Resist Invasion by Invasive Exotics

A diverse healthy native plant community is more resistant to invasion by opportunistic exotic plants.

Why are Aquatic Plants Seen as a Problem?

If aquatic plants are so wonderful, why are they perceived as a problem? Most of the time, problems arise when plants are so numerous they impede recreational activities such as boating and swimming. When growth becomes very thick, they also harm some fisheries, particularly juvenile salmon and trout habitat. The causes of unnaturally high levels of plant growth are complex. Often it is attributed to increased nutrients, which come from around the lake or in the watershed. Contributing problems can include failing septic systems, fertilizer run-off, or agricultural waste. These increased nutrients cause the natural process of lake aging (eutrophication) to proceed at an accelerated rate, and increased plant and algal growth is part of this process.

Another problem can arise if a nonnative species is inadvertently introduced to the lake. This often happens when recreational users unknowingly carry plants from one waterbody to another, or when someone discards aquarium plants into a lake. Several exotic species such as Eurasian watermilfoil (*Myriophyllum spicatum*) or Brazilian elodea (*Egeria densa*) are aggressive and can crowd out more desirable native vegetation. Changes in vegetation may take place slowly or quite rapidly.

Another problem can arise if a nonnative species is inadvertently introduced to the lake. This often happens when recreational users unknowingly carry plants from one waterbody to another, or when someone discards aquarium plants into a lake. Several exotic species such as Eurasian watermilfoil (*Myriophyllum spicatum*) or Brazilian elodea (*Egeria densa*) are aggressive and can crowd out more desirable native vegetation. Changes in vegetation may take place slowly or quite rapidly.

Solution

Along with preventing or eliminating pollution, you can monitor plant community changes by collecting and identifying aquatic plants on a year-to-year basis. This is also a good way to detect detrimental changes at an early stage when control or elimination of the problem is both less complicated and less costly.

Collecting and preserving plants is not difficult, and the result is an increased awareness of aquatic plants, as well as a valuable historic record of what grows in the lake. However, proper identification of the plant can be tricky and is essential, particularly if you believe any to be exotic invasive species.

Benefits and Disadvantages of Aquatic Plants in Ponds

Aquatic plants are not considered desirable by many pond owners as they are viewed as a nuisance to such pond activities as swimming and fishing. However, all aquatic plants should not be viewed as undesirable, given the benefits they can provide. Many plants are critical to fish and wildlife communities associated with ponds because they provide cover, nesting areas, and food.

Ultimately, whether a plant provides a benefit or is considered a nuisance rests entirely with the pond owner and his/her goals for the pond. For example, a pond owner who desires a high-quality swimming pond often views any aquatic plant as undesirable and may go to great expense to eliminate these plants. Conversely, a pond successfully managed for wildlife will have a variety of aquatic plants present with only a few plant species being considered a nuisance.

The purpose of this fact sheet is to provide pond owners with insight into the value or negative aspects associated with various aquatic plant species.

If control of aquatic vegetation is the desired outcome, Ohio State University Extension Fact Sheets A-3-98, Controlling Filamentous Algae in Ponds, and A-4-98, Chemical Control of Aquatic Plants, should be obtained and read carefully. The University of Floridas aquatic plant web site at aquat1.ifas.ufl.edu/photos.html provides excellent pictures of most aquatic plants found in Ohio ponds.

Planktonic Algae

These algae are microscopic in size. Planktonic algae assemblages are commonly composed of diatoms, blue-green algae, and green algae, although other species can be found in Ohio. A better understanding of planktonic algae is available in Ohio State University Fact Sheet A-9-01, Planktonic Algae in Ponds.

Benefits

Planktonic algae are the foundation of the aquatic food chain in all ponds and lakes, and their abundance ultimately determines how many pounds of fish, such as largemouth bass and bluegill, can be grown in a pond.

Algal abundance is determined by water fertility with higher levels of phosphorus and nitrogen resulting in increased abundance. Higher planktonic algae density enhances food production for fish, making the pond capable of producing more pounds of fish.

As planktonic algae abundance increases, light penetration to deeper water becomes more limited. This, in turn, limits how far from shore submerged aquatic plants and filamentous algae can grow and whether or not they can or will become a nuisance for the owner.

As a rule of thumb, water clarity of 24 to 30 inches is desirable for fish production and for limiting other plant species abundances to tolerable levels.

Disadvantages

Too much of a good thing certainly can apply to planktonic algae if nutrient levels are too high. These algae can explode in response to high levels of nutrients and literally can turn a pond pea green, resulting in very low water clarity. In rare instances, pre-dawn oxygen levels can be reduced to lethal levels for aquatic animal life.

The ratio of nitrogen to phosphorus determines whether the planktonic community will be dominated by green algae that are readily eaten by microscopic animals or by blue-green algae that offer less value to the food chain. Low ratios of nitrogen to phosphorus favour blue-green algae rather than the more edible forms of green algae.

Filamentous Algae

Filamentous algae are microscopic algae that form colonies of filaments hence the name. These algae are notorious for forming the large, pillow-like mats of algae that float on the surface of ponds. Common types found in Ohio include Spirogyra and Pithophora.

Benefits: None in recreational ponds.

Disadvantages

As in the case with planktonic algae, high levels of nutrients can cause filamentous algae abundance to explode, especially in ponds lacking other aquatic plants, becoming so abundant that severe oxygen problems can result in the pre-dawn hours during July and August. Treating a severe filamentous algae problem in summer will almost certainly cause a fish kill. Ohio State University Extension Fact Sheet A-8-01, Winter and Summer Fish Kills in Ponds, provides insight into how these types of summer kills occur.

Excessive growth of filamentous algae ruins swimming in many ponds every year. No one enjoys swimming in a pond in which long filaments of algae cling to everything they touch. Additionally, anglers become frustrated with having to remove filamentous algae from their lures after every cast.

Submerged Plants

Submerged aquatic plants are common to most Ohio ponds, unless treatment with a herbicide has occurred or grass carp (white amur) have been stocked. Aquatic plants resemble terrestrial plants in many regards, but lack stem rigidity when removed from the water.

Benefits

Submerged plants are critical to a well-structured fish assemblage. They not only provide protection for small fish from predators but also produce large numbers of invertebrates for small fish, such as bluegill, to eat.

Research has shown that the optimal abundance of submerged plants for largemouth bass-bluegill populations is 15 to 20% of the pondýÿs surface area. This proportion balances the predator-prey relationship between largemouth bass and bluegill so that both species have all sizes of fish represented in the assemblage.

Submerged plants also are an important food source for many species of waterfowl in the form of vegetation-dwelling invertebrates or the plants themselves. Notable duck species associated with submerged plant beds are blue-winged and green-winged teal, wood duck, gadwalls, American widgeon, and northern shoveler as well as several species of grebes. Many species of herons and egrets hunt the shallow areas of ponds for small fish and frogs where submerged plants occur. A pond managed for wildlife needs submerged plants as a habitat component.

Submerged plants also have an effect on water quality. Their ability to put oxygen into the water is an obvious contribution, but they also provide for long-term storage of nutrients that might otherwise be used to create nuisance levels of planktonic or filamentous algae. Ponds with beds of submerged plants have fewer problems with algae.

Disadvantages

As with all aquatic plants, an excessive amount of submerged plants can cause problems for the pond owner. The exception might be for the dedicated wildlife pond in which no other use is desired. When submerged plants are too abundant, they can cause problems similar to those described for filamentous algae.

Excessive amounts of submerged plants can present a problem for the pond owner who values his/her fishery. Once plant levels exceed 20% of the surface area, largemouth bass predation on bluegill becomes less effective. Small bluegill can effectively avoid being eaten by darting into the excessive vegetation. Thus, too many bluegill survive, and their growth declines due to increased competition from overcrowding. Bass growth also decreases, as they are unable to find prey effectively and consume enough to grow well. The classic symptoms of this scenario are populations of small, thin largemouth bass and bluegill. Excessive submerged plants also pose a problem in ponds where water is being pumped out for irrigation, livestock watering, or might be pumped out in case of a fire. These plants can clog a pump intake, lessening the amount of water being pumped and shortening the lifespan of the pump due to excessive wear and tear.

Floating Plants: Floating aquatic plants come in all shapes and sizes from the pinhead size of a watermeal plant to the large leaves of a water lily.

Benefits

In general, watermeal and duckweed provide few benefits to a pond unless the pond is strictly a wildlife pond. The same species of waterfowl that eat submerged plants will also eat these small floating plants. Pond owners who value the natural, aesthetic qualities of a pond desire water lilies and their showy flowers. Largemouth bass and bluegill both like the shade provided by the large leaves of water lilies. The key is not to allow the lilies to overrun the ponds shallow areas, maintaining about 15 to 20% coverage

Disadvantages

As with other aquatic plants, an excess of nutrients can cause an overabundance of watermeal and duckweed. It is not uncommon for these small plants to completely cover a pondýÿs surface in a few short weeks. This type of cover will substantially reduce light penetration so that oxygen-producing photosynthesis in the water underneath the floating plants ceases.

Warm water temperatures during summer months keep oxygen-consuming animal respiration and decomposition rates high, possibly resulting in a fish kill from lack of oxygen.

Water lilies are notorious for overspreading a pond in short order, especially if the pond is shallow and the water is clear. Clear water allows these lilies to send up stems and leaves from a greater depth.

This severely inhibits recreational activities. Many pond owners have planted water lilies for aesthetics, only to be frustrated with their quick coverage of the pond.

Emergent Plants

Emergent aquatic plants are plants that live in shallow water or just on the shore along a pond. Most of the vegetative portion of these plants grows above the water rather than under the water. From a distance, many resemble grasses. Table 2 provides a list of the emergent aquatic plant species common to Ohio ponds.

Benefits

Emergent plants provide important wildlife benefits to ponds. The plants themselves provide nesting habitat for a variety of bird species, such as song sparrow, red-winged blackbird, and wrens. Many other aquatic bird species, such as rails and herons, utilize emergent plant habitats during migration. Many mammals will make a meal of the green vegetative material growing above the water. Additionally, the seed heads of nutsedges, spikerushes, and bulrushes are attractive to many waterfowl species (mentioned previously) as a food source. A pond managed for wildlife should strive to have a variety of emergent plants.

Disadvantages

An overabundance of emergent plants can create a problem for some pond owners. This is particularly true for cattails, which account for nearly all pond owner complaints about emergent plants. Cattails have the ability to completely surround a pond and extend several feet into the water if allowed to do so. Even for a pond managed for wildlife, this poses a problem. Habitat diversity is good for wildlife, and a monoculture of cattails does not provide diversity. Excessive emergent plants can create problems for anglers who will find it difficult to effectively fish from shoreline areas. Cattails are highly attractive to muskrats, a mammal that can damage a pond in some circumstances. Dams are vulnerable to their burrowing activities, and muskrat burrows can compromise their integrity. Muskrat use cattails for a variety of reasons, including food, den material, and as an escape from predators.

One emergent plant species deserves special mention. Purple loosestrife is an invader to Ohio ponds and wetlands and should be controlled, either by pulling the plant and roots or by spraying a herbicide. If left uncontrolled, purple loosestrife quickly spreads and crowds out desirable native emergent plants. A monoculture stand of loosestrife is the result, providing no benefits to the pond owner or wildlife.

Explosion of Aquatic Plants

Massive occurrence of floating aquatic plants is a common feature on tropical and subtropical water bodies, and it is commonly the result of an expansive growth of non-endemic aquatic macrophytes. In a 1989 report to FAO, Hartley (CSIRO, Brisbane, Australia) estimated that two million people in Nigeria, including more than 24 000 fishermen, and approximately 100 000 persons living in riverine communities in Benin, who rely solely on fishing for their livelihood, are negatively affected by floating aquatic weeds (water hyacinth, salvinia, and water lettuce - *Pistia stratiotes*). Fishermen in Malawi and Ghana may be hampered by water lettuce. In the Niger, the more recent explosive growth of water hyacinth has interfered with fishing on floodplains of the Niger River. In lakes Kyoga and Victoria in East Africa, until very recently, thousands of fishermen were trying to keep open access to landing and fishing sites blocked by water hyacinth. Water hyacinth and salvinia are also widespread in the tropical zone of Asia and Latin America, representing a major problem for fisheries, irrigation, hydropower production and water transport.

Both the floating and submersed aquatic macrophytes can become nuisance (noxious) plants, especially where they appear as exotics. The problematic submersed plants are: *Myriophyllum spicatum, Hydrilla verticillata, Ceratophyllum demersum, Potamogeton crispus, Potamogeton pectinatus*. The reason for their explosive growth is often the favourable nutrient environment, and also human impact through eutrophication of some, especially small water bodies. The rapid expansion of the exotic water hyacinth in Sudan reduced the number of fish and fishing intensity (Davies, 1958/59). The periodical annual invasions of the Kerala (India) backwaters with salvinia renders fishing and boating impossible during certain months (Thomas, 1979). In the Sepik River in Papua New Guinea, where in the late 1970s and early 1980s salvinia spread over numerous floodplain lakes, eventually covering over 200 km, the tilapia (*Oreochromis mossambicus*) fishery collapsed (Mitchell *et al.*, 1980). The disruption of traditional subsistence lifestyle became so great that entire villages were abandoned (Anon., 1985). Fortunatelly, biological control using the beetle *Cyrtobagous salviniae*, which was released over the infested areas, was so effective that within only a few years salvinia virtually disappeared from all backwaters. Not always the presence of invasive plants becomes a nuisance. On Rawa Pening reservoir in Java, Indonesia, where all fish species are captured for food, control of water hyacinth led to overfishing, and only expansion of aquatic weeds due to the neglect of control measures led to rehabilitation of fish stocks (Carlander, 1980). Later on, a certain balance has been achieved among the water hyacinth area cover, capture fishery, and fish cage fish production.

The impact of invasive plants on fish stocks and fisheries is poorly documented, especially for the tropical and subtropical countries. Long-term monitoring is mostly lacking. Records available for the freshwater Lake Naivasha in Kenya, are an example of well-documented changes in a lake ecosystem under the impact of changes in aquatic macrophytes, but also of the impact of human interference with this freshwater lake.

Over the last 30 years, Lake Naivasha, situated at an altitude of 1 890 m, has experienced considerable changes in its plant as well as animal communities. *Salvinia molesta* appeared in the early 1960s, and toward the end of the 1960s it covered 2–3 km. Its appearance, combined with heavy fishing pressure on tilapia stocks, resulted in a decline to a level that the local frozen fillet plant had to be closed (Gaudet, 1976). Dissolved oxygen concentration underneath the mats was only 10 % saturation as compared to 64–85 % in the open water (Kongere, 1979). On Lake Kariba, Donnelly (1969) observed that tilapias established nurseries in shallow water only if it was free of *Salvinia*. Where salvinia mats shaded the water, phytoplankton production declined and the young tilapias moved out of the shallows into the deepwater *Potamogeton* and *Ceratophyllum* beds, where they were much more susceptible to predation. In Lake Naivasha the large mat of *Salvinia* almost disappeared by 1987, when it became limited to the river inflow along the northern shore (Harper *et al.*, 1990) Since 1991 the introduction of the biological control agent, the beetle *Cyrtobugous salviniae*, has kept salvinia under control (Abiya, 1996). The Lake Naivasha level fluctuations, with a high level lasting from 1960 to 1976, followed by a drop, which was followed in 1982 by a sharp rise in water level, has had considerable impact on other aquatic plants. In 1983 the water level started to decline again and by 1985 Kenya experienced a severe drought. By 1987 the lake level returned to the state of the 1950s. Papyrus (*Cyperus papyrus*) plants, formerly floating on the surface, became restricted to margins, but water-lilies (*Nymphaea caerulea*) started germinating, and extensive beds of submersed plants such as *Ceratophyllum demersum* and *Potamogeton* spp reappeared. The impact of these changes on fish and crayfish revealed that *Tilapia zillii* and *Oreochromis leucostictus* (both introduced species) showed resistance to the changes in lake level fluctuations and subsequent changes in aquatic plants. Both fish feed largely on detritus, which has always been in abundant supply.

In *T. zillii* macrophytes constituted less than 10% of the total food taken, while there were none eaten by *O. leucostictus*. Decline in tilapia fishery resulted from intensification of the fishery and the use of small-meshed nets. But whenever the lake water level rose, the flooding of terrestrial vegetation and shallows provided good nursery areas and this led to the recovery of fish stocks (Abiya, 1996).

Crayfish *Procambarus clarkii*, introduced in the lake in 1970 and another important component of the lake fishery, is a major consumer of aquatic plants. In the early 1980s it reached a density of about 4 crayfish m^{-2} in the littoral (Muchiri *et al.*, 1995).

The species is known to reduce aquatic vegetation in lakes in the USA at densities over 3 m^{-2} (Magnuson *et al.*, 1975). The crayfish fishery in Lake Naivasha crashed in 1983, this being followed by the recovery of rooted aquatics. This provides a circumstantial evidence that the decline and subsequent recovery of the native rooted submersed and floating aquatic plants was due to the heavy crayfish herbivory until 1983 and its relaxation thereafter (Harper, 1992). By 1987 the crayfish density was only 0.25 ind. m^{-2}.

Eichhornia Crassipes - Water Hyacinth

Water hyacinth is now widely spread throughout tropical and subtropical countries. Gopal and Sharma (1981) reviewed the available information on this plant for the years preceding their publication. They also provided maps on the global water hyacinth distribution to that date.

Rzoska (1976) described the invasion of *Eichhornia crassipes* in the Sudanese White Nile. The upper Nile swamps (the Sudd) have become a 'perennial' centre of infestation for the downstream Nile. In many water bodies of the Nile floodplains water hyacinth has fringed the shores in a belt of variable width, and it has taken over the function of *Pistia*. In fringing plants, up to a width of about 6 m, Bailey and Litterick (1993) found a rich community of aquatic macroinvertebrates inhabiting the root of this floating plant, and Hickley and Bailey (1986, 1987) found rich fish communities in water hyacinth fringes. Similarly, Green *et al.* (1976) found also rich fish fauna in water hyacinth fringes of Lake Lamorgan (Java, Indonesia). However, Davies (1958/59) and Rzoska (1976) noted that fisheries were impeded, specifically in areas inhabited by Shilluks, immediately north of the Sudd in Sudan, and on the Sobat River, also in Sudan. Gopal and Sharma (1981) estimated that in the 1950s about 45 000 t of fish were lost in West Bengal (India) due to the infestation of waters with water hyacinth. Referring to literature, they reported that fish production declined from 905 kg ha^{-1} at zero percent cover to 281 kg ha^{-1} at 25% cover, possibly due to low densities of phytoplankton as a result of shading by water hyacinth, and also the uptake of phosphorus by this plant.

One observation reported suppression of the phytoplankton production underneath 10% and 15% cover to 2.25 g m^{-2} and 1.97 g m^{-2}, respectively, as compared to 2.58 g m^{-2} at 5% cover (Gopal and Sharma, 1981). In experimental ponds in Alabama the production of the phytoplankton-feeding *Oreochromis aureus* was reduced by about 50% at 10–25% water cover by water hyacinth (McVea and Boyd, 1975).

In other situations, thick mats of water hyacinth cause complete depletion of dissolved oxyxgen content and consequently fish kills. The presence of water hyacinth in shallows makes such areas unsuitable for fish spawning. In the Sudd of Sudan, Bishai (1960/61) measured 1.8 mg L^{-1} of dissolved oxygen 30 cm below the mat when there was no water current, and he also reported high CO_2 concentrations, which increase the susceptibility of fish to low dissolved oxygen concentrations. Poi de Neiff and Solis (1994), who measured dissolved oxygen concentrations among the roots of water hyacinth floating islands on the Parana River floodplain in Argentina, recorded a maximum of 2.3 mg L^{-1} within the first 1 m depth, but the usual concentrations were close to 1 mg L^{-1}, declining with depth.

Two Amazon River floodplain fish species, *Colossoma macropomum* and *Schizodon fasciatum*, have shown considerable resilience to hypoxic conditions. Experimental investigations demonstrated that if the hypoxic conditions persist, these species return to the zones of plant cover and remain among the roots of floating plants, where the oxygen availability among the roots seems to be better than in the deeper water. In experiments with *Eichhornia crassipes* and *Pistia stratiotes*, conducted by Jedicke *et al.* (1989), the roots of these plants released oxygen into the water. The input of oxygen beneath a layer of *E. crassipes* was nearly double that beneath an open water surface. The input from water lettuce (*P. stratiotes*), however, was only slightly greater than that determined in open water. Jedicke *et al.* (1989) has pointed out the advantages for the hypoxia-tolerant fish species to remain among the plant roots: reduced predatory pressure; utilization of food supply when other species cannot access it; lower danger to fish from birds and other terrestrial predators, to which fish species displaying emergency respiratory behaviour at the surface of open water zone are exposed. Bailey and Litterick (1993) found markedly less aquatic macroinvertebrates inhabiting the roots of water hyacinth when appearing as small floating mats, as compared with those of the fixed fringe, and attributed this reduction partly to the feeding activity of fish. It is probable that if the plant appears in broken small-sized mats, which allow to maintain dissolved oxygen concentration underneath the mat at reasonably high levels, its is sought by fish to feed on its periphyton and the associated aquatic invertebrates. The plant also provides a habitat to invertebrate hosts and vectors of parasitic diseases such as malaria, schistosomiasis, encephalitis and elephantiasis.

In the 1970s and 1980s in East Africa water hyacinth invaded the lakes Kyoga, Kwania and Victoria. In Lake Victoria water hyacinth was noticed for the first time in 1988, and by 1992 it was already well established (Ogutu-Ohwayo, 1995). In the 1990s it has been recorded also from Lake Albert. It has had a major negative impact on fish catches and fish landings, due to its obstruction of fishing and landing sites. It is also believed that its presence along the shores has had a major impact on tilapia nesting, through lowering the dissolved oxygen above such sites and shading. In the early 1990s the worst affected areas have been on Lake Kyoga (FAO, 1993a), but by 1996, in the Uganda sector of Lake Victoria, the weed extended an average of 15 m from the shore, covering almost 80% of the country's 1 000 km of shoreline. Near Kampala, the capital of Uganda, the weed was 30 m wide. Prevailing winds from the south pushed most of the weeds to Uganda, but in 1996 the weed almost encircled the 3 000 km of shoreline (Anderson, 1996). Luckily, the situation has dramatically changed for the better, and by January 2000 it was reported that water hyacinth had been reduced by 80% on Lake Kyoga and 70% on Lake Victoria (Wendo and Ngatya, 2000). In Lake Victoria only 16% of the shoreline remains infested. This reduction is believed to be the result of the success of biological control using two weevil species, i.e. *Neochetina bruchi* and *N. eichhorniae*, released on Lake Victoria since 1996, and on Lake Kyoga since 1994.

With the decimation of haplochromine stocks by Nile perch (*Lates niloticus*), especially in Lake Victoria, a potential benefit resulting from the presence of water hyacinth has been observed: this floating plant appears to provide cover for haplochromine stocks which have been on the increase since the appearance of water hyacinth in Lake Kyoga. In the reservoir Rawa Pening in Java, Goeltenboth and Kristyanto (1987) observed that water hyacinth is a valuable habitat for fish as long as it does not form extensive solid mats.

Ogutu-Ohwanyo (1995) in his comments on the potential for controlling water hyacinth in Lake Kyoga implies that reduction in the water hyacinth cover might again expose haplochromines to Nile perch predation. The haplochromine-aquatic macrophyte relationships are far from clear. Witte *et al.* (1992) suggested that continuous digging by haplochromines in Butimba Bay, Lake Victoria, could have prevented the settling of rooted plants such as *Nymphaea* spp. With the decline in haplochromines, since the end of the 1980s, the shallows in this bay have been overgrown by *Nymphaea* spp, *Ceratophyllum demersum* and invaded by water hyacinth. Witte *et al.* (1995)

believed that water hyacinth invasion of Lake Victoria threatened haplochromines and other fish. It would be interesting to know the result of the presently rapidly diminishing water hyacinth cover on haplochromines and other littoral fish.

In Ghana, water hyacinth was found for the first time in Tano lagoon, apparently entering this water body from Cote d'Ivoire. More recently fishermen reported a decline in fish catches and it was suggested that this was due to the presence of the water hyacinth. Much of the water hyacinth in Tano lagoon is mobile, being moved by wind or water currents. Holcik (1995), who investigated the situation, thought that the low catches are not due to the decline in fish stocks but rather a result of both the inefficient deployment of nets and of the difficult access to the fishing grounds due to the water hyacinth cover.

In Laguna de Bay, a large lake near Manila, the capital of the Philippines, infestation by water hyacinth has been a major environmental problem in several areas of this lake. Massive proliferation of the plant not only poses a hazard to navigation, but it also leads to poor water quality in this shallow lake, which has considerable fisheries importance. Tilapia and milkfish have been grown in numerous fish pens, the area of which at its maximum expansion reached almost 30 000 ha, representing about one third of the total water surface. Frequent typhoons have been known to drive the floating plants against fish pens (Photo 6), destroying them and releasing the penned fish into the lake. In the late 1980s water hyacinth cover was estimated to be 1 800 ha, and the plant was also present in tributary rivers, creeks and irrigation canals. This resulted in difficult, if not impossible access to fish pens, fishing sites, damage to fish pens, fishing nets and other fishing gear, reduction in fish population due to the low dissolved oxygen levels and increased carbon dioxide concentrations, blockage of water intakes of irrigation pumps, and interference with water transport. The removal of water hyacinth was largely manual, plus using one plant harvester. Monitoring of water hyacinth infestation in Laguna de Bay was done with the help of remote sensing using Landsat imageries (Bina *et al.*, 1978).

Tin-mine lakes, of which about 4 300 exist in Malaysia, are

a consequence of abandoned tinmining pits, which have filled with water. Some of them have been used for aquaculture, and these represent 18.9% of the total area used for freshwater fish pond culture in peninsular Malaysia (Arumugam, 1994). While a total of 55 species of aquatic macrophytes have been identified from these lakes, in such lakes which receive a significant nutrient input, drastic changes in macrophyte and other communities take place. Once fish culture has been abandoned, in those lakes which are used as dumping sites for organic waste, *Eichhornia crassipes* may cover up to 90% of the surface area. Such lakes contain mainly air-breathing fish (*Channa* spp, *Trichogaster pectoralis*) which tolerate low levels of dissolved oxygen underneath the mat of floating vegetation. Lakes lacking the dense cover of floating macrophytes support indigenous fish and tilapias.

The widespread abundance of water hyacinth has attracted the attention of fish feed specialists and fishery managers. Water hyacinth has been tested both as a direct food of fish, and as a fish feed component. For example, in experiments carried out in Java, Indonesia, up to 10% of the feed was either composted or non-composted water hyacinth. It was noted that the feed, fed to *Puntius javanicus* and *Cyprinus carpio*, did not slow down the growth rate of these fish. *Trichogaster pectoralis* has shown a significantly slower growth rate when fed higher levels of substitution (Hutabarat *et al.*, 1986). Numerous other experiments have been performed on the use of water hyacinth as an additive to fish feed, but they are not a subject of this review.

In Indonesia, water hyacinth, together with some other floating plants, is used as a fish attracting device. In shallow lakes, an enclosure made of bamboo poles pushed into the bottom mud, and filled with aquatic plants, is called '*bungka todo*'. Tree branches or brush are often added to the aquatic plants. In Kalimantan, such systems are used on inland lakes of the Mahakam River in the east. On the Kapuas River in the west, water hyacinth is rarely found as it does not grow well in black waters of inland floodplain lakes, therefore any water hyacinth which is found is treasured and used as shade to protect fish in cages from direct sunshine, rather than using it as *bungka todo* (Petr, 1993) (Photo 8). The plants provide the fish with shade and food, especially periphyton and its associated invertebrate fauna. The floating aquatic plant enclosures are regularly fished. It has been suggested that in lakes threatened by overfishing, such as Lake Tempe in Sulawesi, *bungka todo* could be an efficient way of fish stock protection and preservation in areas set aside and designed strictly off-limits for fishery. From them, the unprotected areas of the lake would then be naturally recolonized, at least by some fish species.

Pistia Stratiotes

Pistia stratiotes is a free floating common aquatic plant in tropical Africa, where it occasionally forms dense and large mats. In most situations, such mats are not permanent, and are moved or dispersed by wind. In some water bodies, *Pistia* may become part of sudd. The plant consists of a rosette of leaves covered with hydrofobic hairs extending above the water surface, and of submersed, finely divided roots reaching to a depth of 15–20 cm. The roots usually harbour a rich community of invertebrate fauna, which in turn is preyed upon by fish and aquatic predatory invertebrates, such as dragonfly nymphs.

In Volta Lake (a reservoir in Ghana, West Africa), large floating mats of *Pistia stratiotes* developed soon after the reservoir started filling. It became abundant especially where shelter against wind and waves was provided by floating or stranded trunks of dead trees, and in between dead trees standing in shallows. In places it was knitted together with *Scirpus cubensis*.

The biggest expansion of *Pistia* carpets occurred toward the start of rising water level. After the water level stabilized, *Pistia* became less numerous, with large plants gone, and only small plants remaining. In *Pistia* plants, situated well inside a mat, roots contained between 5 000 and 16 000 m^{-2} aquatic macroinvertebrates (5–35 g m^{-2} wet weight), excluding zooplankton (Petr, 1968).

The values are lower than those for the margins of floating meadows of varzea, where Junk (1973) recorded a maximum of 780 000 individuals (62 g m^{-2}). Junk included the very abundant crustaceans (Copepoda and Cladocera) in his samples. Throughout the observation period, in Lake Volta the *Pistia* roots macroinvertebrate fauna was dominated by the predatory dragonfly nymphs, which although low in numbers exceeded browsers and filter feeders in biomass. The presence of macroinvertebrates was considered to be of importance for fish feeding on periphyton and on it associated invertebrates. In 1965, the first year of the reservoir filling, *Pistia* of Lake Volta also harboured the snail *Bulinus forskalii*, which was very numerous at the peak of the dry season, when the water level was slowly declining and the surface water temperatures were at their highest. In 1966 the first *Bulinus globosus rohlfsi*, a host of schistosome (*Schistosoma haematobium*) parasites, was recorded from the lake (Paperna, 1970) while *B. forskalii* became rare. In one area Paperna collected over 50 *B. t. rohlfsi* from a single *Pistia* plant. When in some areas of the lake *Pistia* started disappearing, the snails thrived particularly on *Ceratophyllum demersum*, which became very common. This change has led to a dramatic increase in the incidence of schistosomiasis among the fisherfolk on this reservoir.

Little is know about the fish feeding on the aquatic fauna and periphyton associated with the submersed parts of *Pistia*. In Lake George, Uganda, experimental catches within and alongside drifting mats of *Pistia*, which entered a bay surrounded by papyrus, haplochromids represented 98.3% of the total number, and 98.2% of the total biomass of fish captured during a 24 hour period (Petr, unpublished). Among the haplochromids *Haplochromis nigripinnis*, *H. aeneocolor*, and *H. angustifrons* represented 66.5, 17.8, and 13.2% by number of the total, respectively, and a biomass in a similar proportion. *H. nigripinnis* is an omnivore feeding predominantly on algae, insect larvae and emerging insects; *H. aeneocolor*, which is abundant along the papyrus edge, feeds on insect larvae and adult aquatic insects, and the guts contain also plant fragments, suggesting that this species may selectively forage on the food among the roots of papyrus and *Pistia* and that the plant fragments are roots taken incidentally with the insects; *H. angustifrons* is predominantly an open lake species feeding on insect larvae, especially chaoborids, and on zooplankton (Dunn, 1975).

Pistia may be used as a shading plant in aquaculture. In tropical Australia, it has been used in a raceway culture of juvenile growout of freshwater crayfish *Astacus quadricarinatus*.

The use of this plant was considered advantageous for several reasons: provision of suitable habitat/substrate for juvenile crayfish; photosynthetic production of oxygen; stabilization of water temperature; absorption of nitrogenous wastes; provision of natural food organisms associated with the plant, i.e. epibionts; shading effect; high carrying capacity for crayfish juveniles with reduced intra-specific competition; enhancing survival and growth of the juveniles (Jones, 1988).

Salvinia Molesta

Exotic aquatic fern, the floating *Salvinia molesta* is spread widely through the tropics. Thick mats of this plant prevent or make difficult boat traffic (Photos 9 and 10), impede or make fishing impossible, and the plants harbour also invertebrate hosts of water-borne diseases.

Solid mats of salvinia depress dissolved oxygen concentration, increase concentration of carbon dioxide, and occasionally may result in formation of hydrogen sulfide in water underneath it. Lake Kariba, a large reservoir on the Zambezi River, was invaded by salvinia shortly after the formation of this reservoir.

After some years salvinia, covering at its largest expansion in 1962 approximately 1 000 km^2, i.e. 23% of the reservoir surface area (Karenge and Kolding, 1995), declined in the 1970s to about 1% (Mitchell and Rose, 1979), largely due to nutrient deficiencies.

Salvinia was replaced by increasing amounts of rooted macrophytes down to a depth of around 10 m. In parallel with that there was an increase in the benthic macroinvertebrates and in benthic-feeding fish species such as *Mormyrus longirostris*, *Synodontis zambezensis*, *Serranochromis codringtonii*, *Serranochromis macrocephalus* and *Labeo altivelis*. Increased diversity within the inshore fish community in Lake Kariba was attributed to the development of a mosaic of new niches among the aquatic macrophytes, periphyton and submersed trees where a variety of benthic gastropods, molluscs, insects and other invertebrates provided a new food resource to fish (Karenge and Kolding, 1995). The lake littoral is, however, not a stable environment due to a drawdown of this reservoir, which serves the production of hydroelectric power for Zambia and Zimbabwe. It was recommended that to promote full development of macrophytes, the drawdown should not exceed 3.5 m. This limit is sometimes exceeded in Lake Kariba, where dry years may irregularly alternate with wet years. In spite of that, the development of a macrophyte belt has enhanced the stocks of juveniles and small fish species such as *S. macrocephalus*, *Brycinus imberi* and *S. zambezensis* which have increased during the 1980s (Karenge and Kolding, 1995). The authors also mentioned that in parallel with the increase of the above fish species there was a decline in small-sized *Hydrocynus vittatus*, *Hippopothamyrus discorhynchus* and *Marcusenius macrolepidotus*. The reason for this decline is not understood.

With respect to aquatic macroinvertebrate populations on *Salvinia molesta* as compared to those on water hyacinth, Arunachalam *et al.* (1980) reported a higher diversity from the first plant, where they found 66 species, than on the roots of water hyacinth (about 30 species) from the freshwater lake Veli in India. The authors do not provide information on the biomass of the macroinvertebrates on these two plants, hence it is difficult to assess which plant is more important for fish as a source of invertebrate prey.

In Lake Kariba, salvinia rootlets were found in stomachs of a great variety of fish species of herbivorous, omnivorous and even piscivorous character. In *Tilapia rendalli* they were found in 19.6% of fish stomachs, but represented only 6.7% of the total food taken (Mitchell, 1976). Salvinia rootlets, detritus or fragments occurred also in: 9.3% *Haplochromis codringtoni*, 11.1% *Haplochromis darlingi*, 19.2% *Pseudocrenilabrus philander*, 29.6%, *Eutropius depressirostris*, 8.1% *Clarias gariepinus*, 24.0% *Synodontis zambezensis*, 38.2% *Micralestes acutidens* (where it formed 19.2% of the total content), 12.3% *Alestes lateralis*, 11.1% *Mormyrus longirostris*, 5.0% *Marcusenius macrolepidotus*, 10.2% *Hippopotamyrus discorhynchus*, 8.1% *Mormyrops deliciosus*. In the majority of fish species it would appear that salvinia rootlets and detritus were taken when feeding on aquatic macroinvertebrates and plankton residing among the rootlets. For example, out of the 19 fish species investigated for food, 15 contained the crustaceans *Cyclestheria* and *Caridina*, 18 contained the nymphs of the mayfly *Povilla adusta*, 14 contained chironomid larvae, 13 contained trichopteran larvae.

In Sri Lanka, salvinia is often removed from water reservoirs manually, but biocontrol using the beetle *Cyrtobagous salviniae* is more efficient.

Cyperus Papyrus

Papyrus rhizomes are typically rooted in loose detritus in a floating mat. Papyrus may occur as free-floating or stranded, or forming fringe swamp. Occasionally, the lake-side edge of the fringe papyrus swamp will float if the lake level permits.

The extent of papyrus swamplands in East and Central Africa is enormous. In southern Sudan the Upper Nile forms large swamps called the Sudd, which covers approximately 40 000 km^2 (Rzoska, 1974). In Uganda, papyrus encircles completely Lake Kyoga, and is well developed in Lake Victoria basin.

In Lake George, Uganda, where it forms large fringing floating swamps, papyrus flourishes in water with a conductivity of 200 ìS, while in Lake Edward (850 ìS) it is restricted to river mouths (of much lower conductivities) on the Uganda shore of the lake (Lock, 1974). Gaudet (1980), in summarizing the significance of the papyrus for the aquatic fringe ecosystem, pointed out that papyrus swamps supply large amounts of fixed nitrogen to tropical lakes and rivers every year thus allowing an increase in animal and plant production at the swamp edge. Papyrus fringes act to moderate changes along a river system by regulating nutrient flow and recycling. For the most part this is accomplished by production of an organic nutrient output at the expense of an inorganic nutrient input. Papyrus plants effectively extract dissolved nutrients from tropical river systems, but such nutrients are later exported as organic particulate matter or adsorbed to particles which are carried into a lake by through-flow (Gaudet, 1979). In practical terms, papyrus swamps may be effective in removal of organic nutrients from sewage effluents into these swamps (Gaudet, 1978), since papyrus seems to be quite tolerant to pollution. The organic nutrients released from the organic particulate matter or adsorbed to particles represent an energy source to a very diverse tropical fauna and flora, which in turn supports a profitable shallow water fish industry in many parts of Africa (Gaudet, 1980). Summarizing the importance of papyrus swamps along the major rivers in Africa, such as Congo and White Nile, Gaudet characterized them as an ecological buffer zone, which moderates changes along the river system by regulating nutrient flow and recycling. For the most part this is accomplished by production of an organic nutrient input at the expense of an inorganic nutrient input. The organic nutrients serve as an energy source to a very diverse tropical fauna and flora, which in turn supports a profitable shallow water fish industry in many parts of Africa. At lake edges papyrus swamps are often sites of intensive fishing and thus a resource worthy conservation.

For air-breathing fish papyrus swamps and other large swampy divides are not likely to be a barrier to dispersal. For water breathers, the use of and dispersal through papyrus swamps is likely to be limited by the availability of oxygen, the open surface area, and efficiency of oxygen uptake in the species (Chapman and Liem, 1995). In a study of a papyrus swamp, which feeds a small river in Uganda, Chapman *et al.* (1998) found peak of dissolved oxygen concentrations associated with biannual flooding, and they suggested that this provides opportunities for fish to use or move through papyrus. However, peak flood values never reached 50% saturation, and thus may pose barriers for some hypoxia intolerant species. At the lake edge of Lake George in Uganda dissolved oxygen showed a rapid decrease moving inside the papyrus belt (Petr, 1973). Chapman *et al.* (1998) reported *Ctenopoma muriei, Clarias liocephalus* and *Nothobranchius* sp. to be found principally in papyrus swamps and other wetlands of the Lake Victoria Basin. Chapman and Liem (1995) found the small cyprinid *Barbus neumayeri* to inhabit the dense interior of papyrus swamps, due to the presence of large gills and the use of aquatic respiration at the air-water interface. Chapman *et al.* (1995) reported that 12 species of East African cichlids are tolerant of low dissolved oxygen concentration and able to use low oxygen habitat along the margins of Lake Victoria as a refugium against the introduced Nile perch.

In lakes with a more pronounced fall in water level, papyrus stands may become drastically reduced. In Lake Naivasha water level declined by just over 4 m between 1980 and 1987, exposing some 6 500 ha of soils, or about 35% of its 1980 area

(Harper *et al.*, 1993). During this decline exposed soils were cleared of vegetation, mostly stranded *Cyperus papyrus* and other emergent plants, and put down to agriculture. Harper *et al.* (op. cit.) noted that fringing papyrus is now voluntarily kept by some landowners to keep a buffer zone, which assists in reducing the input of high nutrient loads into the lake. The increasing demand for water is seen as a direct threat to the lake in the future (Abiya, 1996). For details on water chemistry in papyrus swamps and underneath floating papyrus mats of the fringing zone, and Thompson (1976).

Submersed Aquatic Plants

Myriophyllum spicatum, a native Euroasian plant, spread rapidly across North America in the 20th century, displacing native species in some water bodies and often forming dense stands and surface mats. In Currituck Sound in North Carolina, USA, it was detected for the first time in 1965, when it covered 40 ha. In 1966 it covered 3 200 ha, and by 1974, 32 000 ha were infested (Sheldon, 1997). *M. spicatum* can withstand a broad range of conditions: it can grow in oligotrophic to eutrophic waters, in 0.5 m to 8 m depth, in sand, in organic substrate, under pH ranging from 5.4 to 10, in brackish water, in northern temperate lakes and rivers, and in subtropical water bodies.

In 1978 in Lake Pontchartrain estuary, this macrophyte has become a dominant species, with a lower abundance of the native macrophytes *Vallisneria americana* and *Ruppia maritima* present there. Duffy and Baltz (1998), who in the estuary studied the impact of the invasive *Myriophyllum* on fish, found that the highest fish species diversity was in *Vallisneria*, while the lowest in *Ruppia* and *Myriophyllum*.

The common fish were significantly more abundant in vegetated areas than on adjacent unvegetated substrate, and total abundances were higher in *Myriophyllum* and *Ruppia* than in *Vallisneria*.

The authors suggested that the exotic *Myriophyllum* may not have had a detectable influence on fish assemblages or abundances relative to the native macrophytes because the disturbance of the substrate by waves in more open areas prevent it from growing densely enough to strongly alter the original microhabitat. However, in some other water bodies, *M.*

spicatum may form thick, impenetrable walls of uniform height and leaf form. When the stem density exceeds 250 stem m^{-2} fish start avoiding the plant (Savino and Stein, 1982). Keast (1984) found a lower fish density in *Myriophyllum* beds than in native aquatic macrophytes, and Dvorak and Best (1982) found lower densities of aquatic macroinvertebrates in this plant, when compared to native plants.

Sloey *et al.* (1997), who studied the distribution of aquatic macroinvertebrates in beds of *M. spicatum* in Fish Lake, Wisconsin, found higher invertebrate biomass, density, and taxa richness in the upper foliated sections of plants versus the lower unfoliated sections. The edges of the beds contained higher biomass, density, and taxa richness than the centre of the bed. They did not study the impact of the differential distribution of macroinvertebrates on fish, but other studies confirmed that the structural complexity of plant species and spatial distribution of plants are among the decisive factors determining which and how many fish are frequenting aquatic macrophyte stands.

Hydrilla verticillata is another invasive submersed species, which is present in numerous water bodies of the Americas, Europe, Africa, Asia and Australia. In some water bodies in the USA this plant has led to a considerable increase in fish stocks and fish species of sport fishery importance, while in other water bodies its presence has required a management response to prevent decline in recreational fishery. Apart from grass carp, for which this plants is a preferred food, other fish have also been used for controlling this plant. In Africa, *Tilapia zillii* and *T. rendalli* are known to feed readily on it, in India *Cirrhinus mrigala* and *Etroplus suratensis*, stocked in nursery tanks with *Channa punctatus*, efficiently controlled this plant (Raj, 1976), and also in India, *Tor tor* and *Puntius* spp feed readily on it. Herbivorous turtles in Bangladesh, and manatee in the USA and some Latin American countries also feed on hydrilla.

Menace of Excessive Aquatic Plants

Water hyacinth (*Eichhornia crassipes*) which have been described as the most troublesome weed in the world have been linked to several problems like obstruction to water transportation, micro-habitat for disease vectors, obstruction to fishing and reduction in biodiversity. However, recent studies have also shown that this macrophyte can be used for the production of paper, biogas, fertilizer, fish feed and in the clean-up of polluted environment (phytoremediation). It becomes important to fully harness the potentials of this aquatic macrophyte, which could change its status from a weed to an income-generating plant.

The aquatic macrophyte called water hyacinth (*Eichhornia crassipes* [Mart.] Solms) is not new in the ecological history of man. Infect, it has been popularly described as the most troublesome weed of the world because of its rate of multiplication. It is aquatic in nature, deriving its energy form the sun, storing this energy in a semi-succulent stem and a fibrous network of roots.

The commonest water hyacinth is the South American species-*Eichhornia crassipes*. It is distinctly divided into three (3) parts; a fleshy leaf (leaves) that form the basis of photosynthetic accumulation, a greenish semi-succulent stem and a rather brownish fibrous root network. The latter is usually submerged under the water and forms the propellant part of the weed.

As a crop, the weed is still a mystery to many ecologists by virtue of its rapid multiplication. To say that it breeds like a rat is an under-statement, to compare it to the hydra or *Paramecium* could be an exaggeration. However, the weed has a high multiplication factor. *Eichhornia crassipes* has as much as seven-fold (7) increase or spread in 50 days time. Infact, it has been reported that 2 hyacinth plants can grow into 1,200 offspring in 120 days.

A possible biological explanation to this behaviour lies in the fact that the plant grows usually in places with high solar energy constants, ranging form 450 to 550 watts per metre square of land. Coupled with this is its high photosynthetic fixation efficiency of 1.52%. When compared to typical crops in sub-saharan Africa like maize, with a value of 1.0%, cocoa with value of 0.5%, elephant grass with a value of 0.48% and groundnut with value of 0.29%, one sees immediately that water hyacinth has a rapid growth tendency.

Cultivation reports from some swampy areas of South America have it that the plant has an annual average productivity of about 350-1,700 tons per hectare of wet vegetation. With this high growth rate, waterways are easily clogged up, because "clearing" rate is far below the "multiplication" rate of the plant.

This present "menace" position could turn to be a blessing in the near future as possible utilization are sought for the seaweed in the production of animal feeds, detergents and recently in phytoremediation; the deliberate introduction of aquatic weeds like water hyacinth into polluted water bodies so that the seaweed so introduced will absorb pollutants like nitrate, phosphate and heavy metals and thus, purifies the water.

Scientific Classification of Water Hyacinth

Kingdom	:	Plantae
Division	:	Magnoliphyta
Class	:	Liliopsida
Order	:	Liliales
Family	:	Pontederiaceae
Genus	:	*Eichhornia*

Species

E. crassipes	:	Common water hyacinth
E. azurea	:	Anchored water hyacinth
E. diversifolia	:	Variable leaf water hyacinth
E. paniculata	:	Brazilian water hyacinth

Elemental Composition of Water Hyacinth: Fresh water hyacinth contains about 90% water and about 15-20% solid materials. On a dry weight basis, the weed contains about 25-35% protein-related matter. Specifically, amino acids constitute about 17% of the total protein matter, the rest being amides. This explains why water hyacinth cannot be eaten fresh like most other edible green vegetables like lettuce (the amides are usually toxic matters).

Further analyses show that the carbon content of the dry weed is about 36-40%. Direct carbonylation reactions reveal that the carbonates and nitrates obtained were in the order of 40 to 60% in yield ratios. Thus, water hyacinth has a predominantly cellulosic structure highly impregnated by the amino group directly at the carbonyl structure. The aquatic plant can be conveniently represented as:

Investigations carried out have revealed R to be more of the aliphatic chain. Summarily, therefore, the water hyacinth in elemental composition is about 12.8% nitrogen content, 36-40% carbon, hydrogen about 8% and oxygen about 13-14%.

The aquatic plant is also known to be "carrier" of heavy metals like iron, magnesium and zinc, which suggest why it's being used in photoremediation. There are negligible traces of phosphorus and calcium. These metallic ligands can possibly account for the rest of the structure.

Description of Water Hyacinth: Water hyacinth *(Eichhornia crassipes)* is a member of the pickere/weed family-Pontederiaceae. The plants vary in size from a few centimeters to over a metre in height. The glossy green, leathery leaf blades are up to 20 cm long and 5-15 cm wide and are attached to petioles that are often spongy-inflated. Numerous dark, branched, fibrous roots dangle in the water from the underside of the plant. The inflorescence is a loose terminal spike with showy light-blue to violet flowers (Flowers occasionally white). Each flower has 6 bluish-purple petals joined at the base to form a short tube. One petal bears a yellow spot. The fruit is a three-celled capsule containing many minute, ribbed seeds.

Geographic Distribution: Water hyacinth, "perhaps the world's most troublesome aquatic weed" is a native of tropical South America that has spread to more than 50 countries on five continents and has become a massive problem in waterways in both Africa and Southeast Asia. Its air-filled tissue (aerenchyma) enables it to float and spread rapidly within and between connected water bodies. It reproduces asexually by breaking apart into pieces each of which develops into a separate plant. This results in a rapid increase in biomass and continuous mats of living and decaying water hyacinth up to two metres thick covering the water surface have been reported.

Water hyacinth was introduced to North America in 1884 via the Cotton States Exposition in New Orleans. The plant was displayed in ornamental ponds and distributed as souvenirs to visitors, with the excess dumped into nearby creeks and lakes. The plant soon overtook these lakes and creeks and controlling it became a problem. According to Joyce (1992), these problems led to passage of the River and Harbour act in 1899, authorizing the United States Army Corps of Engineers to maintain navigation channels in these areas. Control efforts included the spraying of sodium arsenite, which poisoned applicators and livestock. In Nigeria, water hyacinth was first noticed in 1985 in Lagos Lagoon. Since then, it has spread to a lot of water bodies in the country, making navigation and fishing an almost impossible task. Water hyacinth is found in a lot of water bodies in the tropics.

Habitat: Water hyacinths grow over a wide variety of wetland types form lakes, streams, ponds, waterways, ditches and backwater areas. Water hyacinths obtain their nutrients directly from the water and have been used in wastewater treatment facilities. They prefer and grow most prolifically in nutrient-enriched waters. New plant populations often form from rooted parent plants and wind movement and current help contribute to their wide distribution. Linked plants form dense rafts in the water and mud. In the Pacific Northwest, water hyacinth is planted outdoors in ponds and aquaria, but it is not considered winter hardy, except under special conditions.

Reproduction: Water hyacinth (*Eichhornia crassipes* [Mart.] Solms) reproduces sexually by seeds and vegetatively by budding and stolon production. Daughter plants sprout from the stolons and doubling times have been reported by Westerdahl and Getsinger (1988) as 6-18 days. The seeds can germinate in a few days or remain dormant for 15-20 years. They usually sink and remain dormant during periods of stress (droughts). Upon reflooding, the seeds often germinate and renew the growth cycle.

Control of Water Hyacinth: There are three common methods that have been used to control water hyacinth. These are:

- Mechanical control method
- Chemical control method
- Biological control method.

Mechanical Method: Ever since it colonised the Northern Rivers of New South Wales with such devastating speed, water hyacinth has tested human ingenuity in devising control techniques. Early control attempts concentrated on removing the plant from water with hand or instrument like pitchforks, then dumping the accumulated mass on land to die. This control method is costly in time, money and energy and several of the procedures used damaged the ecology, affecting all animal life in the aquatic ecosystem infested by the hyacinth.

Chemical Method: This method involved the use of herbicides to control water hyacinth. Various kinds of herbicides such as 2, 4-D, Dalapon, Diquat and others have been used in some places. They also opined that the ecological problems created by these herbicides were obvious. The water could not be used for irrigation or human consumption for long period of time and the fauna in the ecosystem were seriously affected. Raju and Raddy working on some herbicides concluded that the best chemical control was achieved with 2, 4-D dimethyl amine 58% (4 kg ha^{-1}). They also reported that the cost of removal by this herbicide was 61% less than that of manual weeding.

Biological Method: Biological control of water hyacinth has been studied with several kinds of animals like viruses, bacteria and fungi as well as with Manatees, insects, herbivorous fish such as grass carp and Tilipia, ducks, geese; turtles, snails and other animals. However, the results were disappointing, perhaps because of defense mechanisms in the plants. For example, the larger plants form 2.5 or more leaves for each one destroyed by pathogen attack. Two insects that have enjoyed the widest usage in the biological control of water hyacinth are-*Neochetina eichhorniae*-aweevil and *Neochetina bruchi*-a weevil also. A moth; *Sameodes albiguttalis* have also been used. Jayanth (1988) used *Neochetina eichhorniae* and got an impressive result where 95% of the infestation was cleared within 32 months. Another method of controlling water hyacinth (*Eichhornia crassipes*) is integration of mechanical, chemical and biological methods because of the inadequacies of each method. Charudattan, (1986) developed a practical integrated water hyacinth control system based on a pathogen, *Cercospora rodmanii*; arthropods, including *Neochetina eichhorniae*; a moth,

Sameodes albiguttalis and a mite, *Orthogalumna terebrantis* and 2 chemical herbicides; 2, 4-D and diquat. He reported that the pathogen or the arthropods alone did not completely control water hyacinth, but that their combination was highly synergistic, yielding 99% control after 7 months.

The Problems Caused by Water Hyacinth Infestation: Water hyacinth can cause a variety of problems when its rapid mat-like proliferation covers areas of fresh water. Some of the common problems are listed below:

Hindrance to Water Transport: Access to harbours and docking areas can be seriously hindered by mats of water hyacinth. Canals and freshwater rivers can become impassable as they become clogged with densely intertwined carpets of the weed. The Ologe Lagoon and the entire Lagos Lagoon complex have been rendered almost impassable by this aquatic plant.

Clogging of Intakes of Irrigation, Hydropower and Water Supply Systems: Many large hydropower schemes are suffering from the effects of water hyacinth. The Owen Falls hydropower scheme at Jinja on Lake Victoria is a victim of the weeds rapid reproductive rates and an increasing amount of time and money has to be invested in clearing the weed to prevent it entering the turbine and causing damage and power interruptions.

Water hyacinth is now a major problem in some of the world's major dams-the Kariba dam which straddles the Zambia-Zimbabwe border on the Zambezi River and feeds Harare has pronounced infestations of the weed.

Blockage of Canals and Rivers Causing Flooding: If water hyacinth is allowed to grow uncontrolled, it can block canals and rivers, which could lead to flooding.

Micro-habitat for a Variety of Disease Vectors: The diseases associated with the presence of aquatic weeds in tropical developing countries are among those that cause the major public health problems: malaria, schistosomiasis and lymphatic filariasis. Some species of mosquito larvae thrive on the environment created by the presence of aquatic weeds, while the link between schistosomiasis (bilharzia) and aquatic weed presence is well known. Although, the statistical link is not well defined between the presence of aquatic weeds and malaria and schistosomiasis, it can be shown that the brughian type of filariasis (which is responsible for a minor share of lymphatic filariasis in South Asia) is entirely linked to the presence of aquatic weeds.

Increased Evapotranspiration: Various studies have been carried out to ascertain the relationship between aquatic plants and the rate of evapotranspiration compared with evaporation from an open-surfaced water body. Haider (1989) suggested that the rate of water loss due to evapotranspiration can be as much as 1.8 times that of evaporation from the same surface but free of plants. This has great implications where water is already scarce. It is estimated that the flow of water in the Nile could be reduced by up to one tenth due to increased losses in Lake Victoria from water hyacinth.

Problems Related to Fishing: Water hyacinth can present many problems for the fisherman. Access to sites becomes difficult when weed infestation is present, loss of fishing equipment often results when nets or lines become tangled in the root systems of the weed and the result of these problems is more often than not, a reduction in catch and subsequent loss of livelihood. Mailu (2001) reported that information from the Fisheries Department, Kenya indicated that there was a 28% increase in total annual fish catches between 1986-1991 and 1991-1997, from 133,097 to 169,890 tonnes.

There was an increase in all species of fish caught except *Oreochromis, Clarias* and *Mormyrus,* which showed declines of 14, 37 and 59%, respectively, over the same period. These declines may have been associated with the inability of fishermen to access the fishing grounds for these species because of water hyacinth infestation.

Generally therefore, as a result of water hyacinth infestation, accessibility to land and water has been hindered, resulting in reduced fish catches, especially of tilapia and mudfish which are found mainly along the shores. Fisherfolks, however, reported increased fish catches from suitable breeding grounds provided by water hyacinth e.g., tilapia, synodontis, protopterus and labeo (Mailu, 2001).

There is, however, need to clarify this conflicting information; in many more areas around the lake and in other parts of the world. A reduced fish catch would have an adverse effect on the quality of life of the communities around the lake and consequently affect sustainable development in the region.

Reduction of Biodiversity: Infestations of water hyacinth affect biodiversity. Dense mats of the weed covering the water surface lead to deoxygenation of the water, thus affecting all aquatic organisms. Death of water hyacinth mats may influence changes in the composition, distribution and diversity of aquatic organisms as follows:

- Displacement of hydrophytes and depressed algal biomass (Twongo *et al.*, 1995; Twongo and Balirwa, 1995)
- Increase in diversity and abundance of some taxa of macrofauna, especially at the borders of the weed mats (Wanda, 1997)
- Increase in the distribution and abundance of schistosome (bilharzia) snail vectors e.g., *Biomphalaria* sp. and *Bulinus* sp.
- Willoughby *et al.* (1993) reported that, based on studies on the Ugandan shoreline of Lake Victoria, mats significantly depressed the diversity of fish species and fish biomass. It was subsequently demonstrated that fish diversity, particularly small taxa, increased along the edge of water hyacinth mats (Twongo and Balirwa, 1995)

Water Supply: Water supply to both villages and municipalities is affected by water hyacinth. In municipalities, water hyacinth interferes with the water intake points through blockage, which lowers the quantity of water pumped. In Kisumu, the municipality reports that the quantity of water supplied has dropped from 20,000 to 10,000 m^3 per day.

Water hyacinth infestations have been reported to lower the water quality in Kenya and Uganda (in terms of colour, pH, turbidity (suspended solids) of water) and hence increase the treatment costs.

Increased costs are associated with keeping the water intake points free of water hyacinth. For example, Kisumu Municipality employs 12 casuals per day, 6 drivers and 6 boat operators, while Homa Bay municipality engages 2 divers at a cost of 1000 Kenya shillings (Ksh) per day. In Homa Bay municipality, water hyacinth builds up 3 to 4 times in a week and it takes 3-4 h to remove it.

Uses of Water Hyacinth

Paper: The Mennonite Central Committee of Bangladesh has been experimenting with paper production from water hyacinth for some years. They have established two projects that make paper from water hyacinth stems. The water hyacinth fibre alone does not make a particularly good paper but when the fibre is blended with waste paper or jute, the result is good. The pulp is dosed with bleaching powder, calcium carbonate and sodium carbonate before being heated.

The first project is quite large with 120 producers involved in paper manufacture. The equipment for pulping is relatively sophisticated and the end product is of reasonable quality. The second project involves 25-30 people and uses a modified rice mill to produce pulp. The quality of the paper is low and is used for making folders, boxes, etc. Similar small-scale cottage industry papermaking projects have been successful in a number of countries, including the Philippines, Indonesia and India.

Fibre Board: Another application of water hyacinth is the production of fibre boards for a variety of end uses. The House and Building Research Institute in Dhaka has carried out experimental work on the production of fibre boards from water hyacinth fibre and other indigenous materials. They have developed a locally manufactured production plant for producing fibreboard for general-purpose use and also a bituminised board for use as a low cost roofing material. The chopped water hyacinth stalks are reduced by boiling and then washed and beaten. The pulp is bleached and mixed with waste paper pulp and a filter agent such as china clay and the pH is balanced. The boards are floated in a vat on water and then finished in a hand press and hung to dry. The physical properties of the board are sufficiently good for use on indoor partition walls and ceilings.

Yarn and Rope: The fibre from the stems of the water hyacinth plant can be used to make rope. The stalk from the plant is shredded lengthways to expose the fibres and then left to dry for several days. The rope making process is similar to that of jute rope. The finished rope is treated with sodium metabisulphite to prevent it from rotting. In Bangladesh, the rope is used by a local furniture manufacturer who winds the rope around a cane frame to produce an elegant finished product.

Basket Work: In the Philippines, water hyacinth is dried and used to make baskets and matting for domestic use. The key to a good product is to ensure that the stalks are properly dried before being used. If the stalks still contain moisture then this can cause the product to rot quite quickly. In India, water hyacinth is also used to produce similar goods for the tourist industry. Traditional basket making and weaving skills are used.

Charcoal Briquetting: This is an idea which has been proposed in Kenya to deal with the rapidly expanding carpets of water hyacinth which are evident on many parts of Lake Victoria. The proposal is to develop a suitable technology for the briquetting of charcoal dust from the pyrolysis of water hyacinth. It is suggested that a small-scale water hyacinth charcoal briquetting industry could have several beneficial aspects for the lakeside communities:

- Providing an alternative income
- Providing an alternative source of biomass
- Improvement of the lake shore environment through the removal of water hyacinth
- Improved access to the lake and less risk to maritime transport
- Reduced health risk associated with the presence of water hyacinth
- Alleviation of pressure on other biomass fuel sources, such as wood, thereby reducing deforestation and associated soil erosion.

The technical aspects are yet to be fully developed and tested but 7 main stages have been identified in the process of converting the plant into charcoal briquettes:

- Harvesting and collection of the plant
- Drying
- Collection and transport to the kiln

- Pyrolysis
- Mixing of the resultant dust with a binder
- Pressing into briquettes
- Marketing of briquettes.

Biogas Production: The possibility of converting water hyacinth to biogas has been an area of major interest for many years. Conversion of other organic matter, usually animal or human waste to biogas is a well established small and medium scale technology in a number of developing countries, notably in China and India. The process is one of anaerobic digestion which takes place in a reactor or digester (an air tight container usually sited below ground) and the usable product is methane gas which can be used as a fuel for cooking, lighting or for powering an engine to provide shaft power. The residue from the digestion process provides an organic fertilizer rich in nutrients.

The use of water hyacinth for digestion in a traditional digester presents some problems. Water hyacinth has very high water content and therefore harvesting effort yields a low reward in terms of organic matter for conversion to biogas. The digester size has to be large compared with that of a traditional type due to the low gas production to plant volume ratio and this can in turn present problems for obtaining an airtight seal. Water hyacinth has to be pre-treated before entering the digester (macerated, chopped or beaten) to promote digestion and to remove air entrapped in the tissue of the plant which would cause it to float. To reduce the need for large volume digesters, high rate digestion techniques have been employed.

One such design has been tested in Bangladesh by a team from Warwick University, UK and the Housing and Building Research Institute, Dhaka, Bangladesh. The design was for a small (8.3 cm) baffled reactor which was fed with juiced water hyacinth. The throughflow of the reactor was 1.2 cubic metres per day. Some cow dung and rumen (taken from a cow's stomach) was added to the water hyacinth juice to promote digestion. Gas was produced in reasonable quantities but some problems were experienced with throughflow and further development is still required. Other studies have been carried out, primarily in India with quantities of up to 4000 l of gas per tonne of semi dried water hyacinth being produced with a methane content of up to 64%. Most of the experiments have used a mixture of animal waste and water hyacinth. There is still no firm consensus on the design of an appropriate water hyacinth biogas digester.

Water Purification: Water hyacinth can be used to aid the process of water purification either for drinking water or for liquid effluent from sewage systems. In a drinking water treatment plant, water hyacinth has been used as part of the pretreatment purification step. Clean, healthy plants have been incorporated into water clarifiers and help with the removal of small flocs that remain after initial coagulation and floc removal or settling. The result is a significant decrease in turbidity due to the removal of flocs and also slight reduction in organic matter in the water. In sewage systems, the root structures of water hyacinth (and other aquatic plants) provide a suitable environment for aerobic bacteria to function. Aerobic bacteria feed on nutrients and produce inorganic compounds which in turn provide food for the plants. The plants grow quickly and can be harvested to provide rich and valuable compost. Water hyacinth has also been used for the removal or reduction of nutrients, heavy metals, organic compounds and pathogens from water.

Animal Fodder: Studies have shown that the nutrients in water hyacinth are available to ruminants. In Southeast Asia, some non-ruminant animals are fed rations containing water hyacinth. In China, pig farmers boil chopped water hyacinth with vegetable waste, rice bran, copra cake and salt to make a suitable feed. In Malaysia, fresh water hyacinth is cooked with rice bran and fishmeal and mixed with copra meal as feed for pigs, ducks and pond fish. Similar practices are much used in Indonesia, the Philippines and Thailand). The high water and mineral content mean that it is not suited to all animals. The use of water hyacinth for animal feed in developing countries could help solve some of the nutritional problems that exist in these countries. Animal feed is often in short supply and although humans cannot eat water hyacinth directly, they can feed it to cattle and other animals which can convert the nutrient into useful food products for human consumption.

Fertilisers: Water hyacinth can be used on the land either as a green manure or as compost. As a green manure, it can be either ploughed into the ground or used as mulch. The plant is ideal for composting. After removing the plant from the water, it can be left to dry for a few days before being mixed with ash, soil and some animal manure. Microbial decomposition breaks down the fats, lipids, proteins, sugars and starches. The mixture can be left in piles to compost, the warmer climate of tropical countries accelerating the process and produces rich pathogen-free compost which can be applied directly to the soil. The compost increases soil fertility and crop yield and generally improves the quality of the soil. Compost can be made on a large or small scale and is well suited to labour intensive, low capital production. In developing countries where inorganic fertilizer is expensive, it is an elegant solution to the problem of water hyacinth proliferation and also poor soil quality. In Sri Lanka, water hyacinth is mixed with organic municipal waste, ash and soil, composted and sold to local farmers and market gardeners.

Fish Feed: The Chinese grass carp is a fast growing fish which eats aquatic plants. It grows at a tremendous rate and reaches sizes of up to 32 kg). It is an edible fish with a tasty white meat. It will eat submerged or floating plants and also bank grasses. The fish can be used for weed control and will eat up to 18-40% of its own body weight in a single day). Other fish such as the tilapia, silver carp and the silver dollar fish are all aquatic and can be used to control aquatic weeds. The manatee or sea cow has also been suggested as another herbivore which could be used for aquatic weed control. Water hyacinth has also been used indirectly to feed fish. Dehydrated water hyacinth has been added to the diet of channel catfish fingerlings to increase their growth. It has also been noted that decay of water hyacinth after chemical control releases nutrients which promote the growth of phytoplankton with subsequent increases in fish yield.

Phytoremediation of Heavy Metals and Crude Oil-polluted Site: Several authors) have reported on the ability of water hyacinth to absorb heavy metals and rid the aquatic environment of these pollutants. The results are quite encouraging. As a matter of fact, reported on passive phytoremediation of heavy metal in Ologe Lagoon, Lagos, Nigeria by water hyacinth. They reported that water hyacinth that was not deliberately introduced into the lagoon to absorb heavy metals did so even when the concentration of the heavy metals in the water column was very small. However, the study on phytoremediation of petroleum hydrocarbon by water hyacinth is still at the infancy. Few studies have been done on it which indicates that water hyacinth can absorb crude oil but more still needs to be done to fully establish the efficacy of water hyacinth in phytoremediation of petroleum hydrocarbon.

Chapter 3

Propagation, Dispersal and Kinds of Aquatic Plants

Of the angiosperms with diverse flowering systems, most have the ability to reproduce clonally, i.e. to produce genetically identical progeny either by vegetative propagation or by forming seeds asexually. Compared with sexually produced seeds, clonally reproduced offspring usually have a limited dispersal ability but a reduced mortality risk, and the propagation of ramets within populations likely to prevent successful seedling recruitment.

Thus, clonal reproduction strongly influences genetic variation, effective population size and metapopulation dynamics. Extensive reliance on clonal reproduction with rare to sporadic sexual reproduction is recognized as a striking convergence in aquatic angiosperms, a biological group that shares several attributes of adaptation to the aquatic condition. For example, some forms of clonal offspring such as turions and shoot fragments are highly effective, economical means of numerical increase, resource acquisition and dispersal under aquatic conditions, reducing the selective value of sexual reproduction;. It may be reasonable, therefore, to consider whether aquatic plants have lower genetic variation or a smaller effective population size than terrestrial plants.

However, the importance of sexual vs. clonal reproduction and their relationship to genetic variation have rarely been studied in aquatic plants, especially floating-leaved, submerged or free-floating taxa. Limited or absent sexual reproduction can result from both ecological and genetic factors. Transient infertility, for example, may arise at the population level because of environmental suppression of seed maturation and dominance of a few genotypes in self-incompatible species. Permanent genetic sterility may be caused by meiotic irregularities associated with hybridization and changes in ploidy level and the accumulation of sterility genes. Population genetic consequences of reproductive traits, therefore, should be clarified in terms of the causes and levels of sterility and its relationship to clonal reproduction.

Utricularia australis R. Br (Lentibulariaceae) is a carnivorous, free-floating aquatic bladderwort widely distributed in temperate and tropical regions, except North and South America. In spite of its widespread distribution, this taxon displays almost complete sterility. A fertile group has been observed only in Japan, where *U. australis* is classified into two forma: sterile *U. australis* f. *australis* and fertile *U. australis* f. *tenuicaulis*. While the phylogeny of *Utricularia* has been studied recently based on cpDNA sequences, the taxonomic and phylogenetic relationships between *U. australis* and its two forma distributed in Japan, *U. australis* f. *australis* and *U. australis* f. *tenuicaulis*, remain unclear. However, it has been demonstrated that (*a*) *U. australis* f. *australis* is a diploid hybrid that originated by asymmetric hybridization between *U. australis* f. *tenuicaulis* (usually as the female parent) and its close relative *U. macrorhiza* (usually as the male parent); (*b*) the hybrid *U. australis* f. *australis* is almost completely sterile, in which only 0.6% of the pollen grains can germinate and almost no seeds can be produced by intra- and interspecies crossing; and (*c*) the absence of post-F_1 generation in natural hybrid populations is confirmed by the additive patterns of amplified fragment length polymorphism (AFLP) bands (Kameyama *et al.*, 2005). It should be noted that AFLP bands common to the hybrid were invariably derived from either parental species, and there are no other parental candidates based on the morphological classification. Interestingly, co-existence of the hybrids and parental species has not been observed.

In the present study, several aspects were addressed: first, whether the capacity of the two parental species, *U. australis* f. *tenuicaulis* and *U. macrorhiza*, to reproduce sexually contributes to higher genotypic diversity compared with the sterile F_1 hybrid, *U. australis* f. *australis*. Propagation of the sterile F_1 hybrids depends solely on clonal offspring in the form of many turions and shoot fragments, while the two parental species use both these forms of clonal reproduction as well as sexual reproduction via seed. Thus, it is reasonable to expect that populations with the capacity for sexual reproduction (parental species) will contain more genotypic diversity than those without (sterile hybrid). Secondly, confirmation is needed of the lack of co-existence of three aquatic bladderworts by exhaustive observation and sampling, because, in a previous study (Kameyama *et al.*, 2005), sample size was limited (only 2–6 samples from each population). Thirdly, an investigation was conducted to address where and how hybrids originate based on the distribution patterns of three aquatic bladderworts and their genetic relationship. The possible scenarios are (*a*) ongoing hybridization in some populations (three aquatic bladderworts that co-exist in some populations have compatible genotypes); (*b*) ongoing but rare occurrence of hybridization combined with long-distance dispersal of offspring (three aquatic bladderworts that are distributed in distant populations are genetically compatible); and (*c*) past hybridization and long-term clonal perpetuation of the hybrids (extant genotypes of three aquatic bladderworts are mutually incompatible).

Materials and Methods

Study Site and Sample Collection

The study was performed in the Tomakomai and Tsugaru regions of northern Japan, which are approx. 250km apart. In the Tomakomai region, the presence and absence of three *Utricularia* had been surveyed over 10 years by Masahiro Toyama, Hokkaido Prefecture, Japan (Y. Kameyama, Hokkaido University, Japan, pers. comm.). In the Tsugaru region, the distributions of three *Utricularia* had been thoroughly surveyed in 1999 and 2000, and two reports were available (Toyama and Katsumata, 1999; Uematsu *et al.*, 2000). A further field survey was conducted in almost all lakes and ponds in both regions. Plant material was collected from all populations where *Utricularia* was observed: 15 populations in the Tomakomai region and 18 populations in the Tsugaru region. An inflatable boat was used to gain access to the whole surface of each water body. Sample sizes ranged from three to 43 (mean = 12; 397 total), with large populations sampled most intensively. Stems of all samples were cleaned with deionized water and frozen at “80°C for later DNA extraction.

AFLP Analysis

Total genomic DNA was isolated from about 50mg of tissue from each of the 397 frozen stems by the CTAB (cetyl trimethyl ammonium bromide) miniprep procedure (Stewart and Via, 1993). An AFLP analysis was performed according to the method of Vos *et al.* (1995), with some modifications. Genomic DNA (approx. 0.1ìg per sample) was digested with the restriction enzymes *Eco*RI and *Mse*I at 37°C for 1.5h. Double-stranded adaptors were ligated to the ends of the digested DNA fragments at 20°C, overnight. Pre-selective amplifications were performed using a primer pair with one additional nucleotide, *Mse*I-C/*Eco*RI-A. Selective amplifications were conducted with three fluorescence-labelled primer combinations, *Mse*I-CC and *Eco*RI-ACT (FAM), *Mse*I-CA and *Eco*RI-ACG (VIC), and *Mse*I-CC and *Eco*RI-AGC (NED). The AFLP Amplification Core Mix (Applied Biosystems, Foster City, CA, USA) and the GeneAmp PCR system 9700 thermal cycler (Applied Biosystems) were used for both amplifications. AFLP fragments were detected with an ABI Prism 3100 automated sequencer (Applied Biosystems) and GENESCAN analysis software (Applied Biosystems).

Statistical Analysis

The number of genotypes observed (G) and the genotypic diversity (D) within populations were calculated for each taxon. Genotypic diversity was estimated as $D = 1$ “ Ó$[n_i(n_i$ “ $1)/n(n$ “ $1)]$, where n is the number of ramets sampled and n_i is the number of ramets with genotype i. The value of D ranges from 0, when all ramets sampled have the same genotype, to 1, when each ramet sampled has a different genotype (Pielou, 1969). Rank abundances of genotypes were calculated for each taxon on the basis of the number of populations in which the genotype was observed. Principal co-ordinate analysis (PCoA) was conducted to reveal the relationship between genotypes. Genetic similarity was calculated as $S_{ij} = 2N_{ij}/(N_i + N_j)$, where N_{ij} is the number of shared bands between genotypes i and j, and N_i and

N_j are the number of bands found in genotypes *i* and *j*, respectively (Dice, 1945). Genetic similarity was transformed to dissimilarity using the formula $D_{ij} = 1$ " S_{ij}, and then subjected to PCoA. All calculations were performed with R Package 4.0 software (Casgrain and Legendre, 1999). *Utricularia australis* f. *australis* is a sterile F_1 hybrid between *U. australis* f. *tenuicaulis* and *U. macrorhiza*, in which no post-F_1 generations are produced (Kameyama *et al.*, 2005). Thus, genotypic similarity between the hybrid and each of the two parental species can be estimated selecting AFLP bands polymorphic in both target taxa, but completely absent in the other parental species. Twenty-two bands were selected for the comparison of the hybrid with *U. australis* f. *tenuicaulis*, and 17 bands for the comparison of the hybrid with *U. macrorhiza*. Genetic similarity (S_{ij}) (Dice, 1945) was calculated as explained above, and an UPGMA (unweighted pair group method with arithmetic mean) cluster analysis was conducted with R Package 4.0 software (Casgrain and Legendre, 1999). Bootstrap values for each branch were determined from 1000 replicates using WinBoot software (Yap and Nelson, 1996). Neighbour-joining trees (Saitou and Nei, 1987) constructed with T-Rex online software (Makarenkov, 2005) confirmed the topologies found with UPGMA except for small differences in the branch length (data not shown).

Results

AFLP Polymorphism

In total, 262 bands were observed in the AFLP analysis, with 209 polymorphic bands. Many species-specific bands were observed in both parental species: 51 for *U. australis* f. *tenuicaulis* and 40 for *U. macrorhiza*. Genotypes of the sterile F_1 hybrid, *U. australis* f. *australis* had most of these species-specific bands: 43–46 of the 51 (84–90%, mean: 88.5%) *U. australis* f. *tenuicaulis*-specific bands, and 33–34 of the 40 (80–85%, mean: 83.1%) *U. macrorhiza*-specific bands. AFLP bands common to the hybrid were invariably derived from either parental species. However, several bands observed in the hybrid were absent in both parental species (7–10 for each genotype, 12 total). The number of polymorphic bands observed within each taxon was 64, 82 and 45 for *U. australis* f. *tenuicaulis*, *U. australis* f. *australi* and *U. macrorhiza*, respectively.

Genotypic Diversity

Seven genotypes were observed in ten populations of *U. australis* f. *tenuicaulis* (T1–T7), nine in 18 populations of sterile F_1 hybrid *U. australis* f. *australis* (A1–A9) and six in seven populations of *U. macrorhiza* (M1–M6) (Table 1). Most populations consisted of a single genotype regardless of the potential fertility of the three taxa:

the mean number of genotypes within populations was 1.1 (s.e. = 0.10), 1.2 (s.e. = 0.09) and 1.1 (s.e. = 0.14) for *U. australis* f. *tenuicaulis*, *U. australis* f. *australis* and *U. macrorhiza*, respectively. Mean genotypic diversity (*D*), including monoclonal populations, was 0.05 (s.e. = 0.05), 0.04 (s.e. = 0.03) and 0.02 (s.e. = 0.02) for *U. australis* f. *tenuicaulis*, *U. australis* f. *australis* and *U. macrorhiza*, respectively.

Genetic Structure

The distributions of the three aquatic bladderworts are mutually exclusive. Only two co-existing populations of the F_1 hybrid, *U. australis* f. *australis*, and either parental species were observed: population 14 (with *U. australis* f. *tenuicaulis*) and population 26 (with *U. macrorhiza*). It is noteworthy that the two parental species never co-existed.

An identical genotype was occasionally observed in multiple populations, even populations distant from one another. In the extreme case, for example, a single genotype (A3) was observed in about half (8/18 = 44%) of the hybrid populations across the two regions. The number of genotypes observed in multiple populations, however, was relatively small: two of seven genotypes for *U. australis* f. *tenuicaulis*, three of nine genotypes for *U. australis* f. *australis* and two of six genotypes for *U. macrorhiza*, with many other genotypes found in only a single population.

Genetic Similarity

Twenty-two genotypes were analysed by PCoA using 209 polymorphic AFLP bands. Along the first axis, genotypes of the sterile F_1 hybrid *U. australis* f. *australis* (A) were distributed midway between those of the two parental species, *U. australis* f. *tenuicaulis* (T) and *U. macrorhiza* (M), consistent with the results of Kameyama *et al.* (2005). Within each taxon, severa genetic groups were found along the second and third axes: T1–T5 and T6–T7 from *U. australis* f. *tenuicaulis*; A1–A2, A3–A5 and A6–A9 from *U. australis* f. *australis*; and M1–M2, M3–M4 and M5–M6 from *U. macrorhiza*. Genetic similarities within each taxon, however, did not correspond to regions. For example, in *U. australis* f. *tenuicaulis*, the genotype group T4–T5 in the Tsugaru region was very different from T6–T7 in the same region.

Principal co-ordinates analysis of 22 genotypes estimated by using 209 AFLP bands from the sterile F_1 hybrid *Utricularia australis* f. *australis* (A) and the two parental species, *U. australis* f. *tenuicaulis* (T) and *U. macrorhiza* (M). The proportion of ...

Although only a small number of AFLP bands were available with which to construct the UPGMA dendrogram, the genetic groups observed were highly consistent with the results of PCoA. One genetic group of the hybrid (A1–A2) was extremely different from all genotypes of both parental species. Genetic group A3–A5 had no similar genotypes in *U. australis* f. *tenuicaulis*, but one genotype of *U. macrorhiza* (M4) was relatively similar. Although genetic group A6–A9 was relatively similar to genotypes in both *U. australis* f. *tenuicaulis* (T6–T7) and *U. macrorhiza* (M1–M2), most of these genotypes were distributed in different regions: for example, T6–T7 in the Tsugaru region and A6–A9 in the Tomakomai region. Moreover, even when the hybrid and a parental species were found in the same locality, genotypic similarities were extremely low: T2–T3 and A7 in population 14, and M2 and A3 in population 26.

Low Genotypic Diversity within Populations

The sterile F_1 hybrid, *U. australis* f. *australis*, showed extremely low genotypic diversity within populations compared with other clonal plants studied to date: the mean number of genotypes per population (*G*) was 1.2 and the mean genotypic diversity (*D*), including monoclonal populations, was 0.04. Even in the parental species, with the capacity to reproduce sexually, the mean number of genotypes per population (*G*) was 1.1 for both species, and the mean genotypic diversity (*D*), including monoclonal populations, was 0.05 and 0.02, for *U. australis* f. *tenuicaulis* and *U. macrorhiza*, respectively. Thus, it is concluded that the capacity to reproduce sexually in aquatic bladderworts plays no role in increasing genotypic diversity within populations.

It is generally expected that populations with the capacity for sexual reproduction will contain more genotypic diversity than those without. However, the relative importance of sexual vs. clonal reproduction may vary among clonal plants, and the extensive reliance on clonal reproduction with rare to sporadic sexual reproduction is recognized as a striking convergence in aquatic angiosperms. In addition, even if many seeds are produced, the seedling establishment largely depends on the existence of safe sites along with the competition with clonal offspring during recruitment. Population genetic consequences of reproductive traits, therefore, should be clarified in terms of not only the capacity for sexual reproduction but also its relationship to clonal reproduction, while only few studies have examined these questions.

The emergent aquatic plant, *Butomus umbellatus*, has marked variation in sexual fertility related to ploidy: diploids produce abundant viable seeds combined with hundreds of tiny clonal bulbils, whereas triploids are sexually sterile and exhibit only limited clonal multiplication through rhizome fragmentation. Random amplified polymorphic DNA (RAPDs) analysis detected one extremely common and widespread genotype for both fertile and sterile populations. More importantly, sexually fertile populations did not exhibit higher genotypic diversity than sterile populations, possibly because seeds are outcompeted by bulbils for safe sites during establishment. Clonal offspring of free-floating aquatic plants, such as turions and shoot fragments, are highly effective for both numerical increase and expansion within populations. In addition, carnivorous, free-floating aquatic bladdderworts can acquire resources solely by shoot fragments, which probably ensure the predominance of clonal reproduction. The rapid expansion of a few founder genotypes via extensive clonal reproduction prevents the subsequent establishment of seedlings and clonal offspring. Competition between clones and stochastic loss of genets may further reduce the genotypic diversity within populations. Widespread expansion of a few clones also inhibits sexual reproduction in self-incompatible species, because it enhances self-pollination or geitonogamy among ramets of the same clone. In aquatic bladderworts, fruits from self-pollination produce fewer seeds than those from cross-pollination: the mean number of seeds per fruit in outcrossed and selfed fruit is 48 and 16 in *U. australis* f. *tenuicaulis*, and 28 and five in *U. macrorhiza*, respectively. Thus, these data suggest that rapid and widespread expansion of a few founders excludes the subsequent establishment of seedlings and clonal offspring, and, furthermore, inhibits sexual reproduction because of reduced seed production caused by increased selfing.

Distribution Patterns of Three Aquatic Bladderworts

Kameyama *et al.* (2005) clarified the fact that *U. australis* f. *australis* is a sterile F_1 hybrid derived from hybridization between two parental species, *U. australis* f. *tenuicaulis* and *U. macrorhiza*. In the present study, the additive patterns of the AFLP bands and the genetic intermediacy of *U. australis* f. *australis* genotypes confirm the hybrid origins and F_1 dominance of this taxon.

In addition, an exhaustive sampling from almost all lakes and ponds in two regions confirmed that the hybrid rarely co-exists with either parental species, and the co-existence of two parental species has not been observed.

Interspecific hybrids are highly variable in fertility and vigor, but, in general, F_1 hybrids of closely related species tend to exceed their parents in vegetative vigor and robustness.

This phenomenon, referred to as hybrid vigor or heterosis, has major implications for evolutionary biology and at least partly explains the success of allopolyploid species and many clonal hybrid lineages. Extreme vegetative vigor has been reported in several hydrophyte hybrids, where hybrids compete with or even displace parental species.

The establishment of *U. australis* f. *australis*, however, may be ensured not only by clonal perpetuation but also by long-distance dispersal of clonal offspring: turions and shoot fragments. Long-distance dispersal by waterbirds is a well known characteristic of aquatic organisms, and the existence of several identical genotypes in multiple and distant populations of sterile F_1 hybrid supports this observation.

Origin of **Utricularia Australis f. Australis**

The observed genotypes of the sterile F_1 hybrid could not have originated from hybridization between the extant parental genotypes within the study regions. This is because (*a*) several AFLP bands observed in the hybrid (7–10 for each genotype) are absent in both parental genotypes; and (*b*) parent and hybrid genotypes in the same region do not show greater genetic similarity than those in distant regions. Thus, possible explanations for the origin of the hybrid genotypes are (*a*) long-distance dispersal from outside the study regions; and (*b*) clonal perpetuation of genotypes that originated from past hybridization event(s).

First, it is considered whether long-distance dispersal from outside the study regions is a possible explanation. In Japan, *U. australis* f. *tenuicaulis* is distributed widely, found almost everywhere except northernmost Japan, but *U. macrorhiza* is found only in northern Japan (Komiya *et al.*, 1997; Y. Kameyama, Hokkaido University, Japan, unpubl. res.). The sterile F_1 hybrid is distributed mainly in northern Japan, in the region where the distribution ranges of the two parental species are broadly overlapping, and only a small number of populations are found in southern Japan (Kadono, 1994; Y. Kameyama, Hokkaido University, Japan, unpubl. res.). Thus, candidate sites for the hybridization between the two parental species are limited to certain regions in Japan near the study regions. Even though clonal offspring of aquatic bladderworts have high dispersal ability via waterbirds (as was discussed in the previous section), it is unlikely that every population of the hybrid was derived from colonizing propagules outside the study regions. Thus, the most plausible explanation for the origin of the sterile F_1 hybrid is that extant populations of this taxon stem from the past hybridization event(s). Long-term perpetuation of clones is possible in clonal plants, even if they are sexually sterile. This scenario can explain why a small number of hybrid populations are found in southern Japan, where one of the two parental species, *U. macrorhiza*, is not found. In addition, hybrids in southern Japan occasionally co-exist with species known as relicts of the last glacial period (Y. Kameyama, Hokkaido University, Japan, unpubl. res.), and both the Tsugaru and Tomakomai regions were greatly influenced by marine transgressions and regressions during that time. Such past climate change could have led to hybridization between parental species that do not co-exist at the present time. However, more thorough research is required to estimate the time and number of hybridization event(s), because climate change is not the exclusive factor to cause the natural hybridization.

The accurate time and number of hybridization event(s) remain unclear. However, it is clear from the present studies that (*a*) *U. australis* f. *australis* originated from past hybridization event(s); and (*b*) hybrid populations have been maintained exclusively by clonal propagation, which may be ensured by both hybrid vigor and long-distance dispersal of clonal offspring. Dispersal is the process of organisms spreading from one place to another. All species, including plants, maintain or expand their range through dispersal. Dispersal is necessary because new generations cannot easily occupy the same physical space their parents occupy.

Unlike most animals, plants have limited abilities to seek out favourable environments. Consequently, plants have evolved adaptations for dispersing by means of seeds, spores or vegetative outgrowths. Plants overcome their little capacity to find favourable environments by producing lots of seeds, ensuring enough of them will settle at favourable sites. Weeds have small seeds, which are better for long-distance travel. Small seeds, however, contain only a limited food supply for the germinating seedling. So there are trade offs between dispersal distance and the food supply. Larger seeds generally have shorter dispersal distances because of their weight.

Plants have evolved dispersal mechanisms that take advantage of various forms of kinetic energy, including gravity, wind, the flow of water and the movement of animals. There is also ballistic/mechanical dispersal, where a seed pod explodes open and flings its seeds away from the mother plant. Read on and you'll learn all about it!

Dis-per-sal by Water

Aquatic and wetland species sometimes use water to disperse their fruits. The coconut for example, produces a large, dry, fibre-filled fruit capable of surviving adrift at least for long periods because of its air pockets. Man-grove seeds germinate while attached to the adult plant and seedlings float away on the sur-face of the water.

The largest seed in the plant king-dom comes from the cocodemer palm, native to the Sey-chelles Archipelago in the Indian Ocean about 1,000 miles East of Kenya.

Algae

Algae are a very large and diverse group of simple, typically autotrophic organisms, ranging from unicellular to multicellular forms, such as the giant kelps that grow to 65 meters in length. Most are photosynthetic like plants, and "simple" because they lack the many distinct cell and organ types found in land plants. The largest and most complex marine forms are called seaweeds.

Though the prokaryotic cyanobacteria are informally referred to as blue-green algae, this usage is incorrect since they are regarded as bacteria. The term *algae* is now restricted to eukaryotic organisms. All true algae therefore have a nucleus enclosed within a membrane and plastids bound in one or more membranes. Algae constitute a paraphyletic and polyphyletic group, as they do not include all the descendants of the last universal ancestor nor do they all descend from a common algal ancestor, although their plastids seem to have a single origin. Diatoms are also examples of algae.

Algae exhibit a wide range of reproductive strategies, from simple, asexual cell division to complex forms of sexual reproduction.

Algae lack the various structures that characterize land plants, such as the leaf-like phyllids of bryophytes, rhizoids in nonvascular plants and the roots, leaves and other organs that are found in tracheophytes (vascular plants). Many are phototrophic, although some groups contain members that are mixotrophic, deriving energy both from photosynthesis and uptake of organic carbon either by osmotrophy, myzotrophy, or phagotrophy. Some unicellular species rely entirely on external energy sources and have limited or no photosynthetic apparatus.

Nearly all algae have photosynthetic machinery ultimately derived from cyanobacteria, and so produce oxygen as a byproduct of photosynthesis, unlike other photosynthetic bacteria such as purple and green sulfur bacteria. Fossilized filamentous algae from the Vindhya basin have been dated back to 1.6 to 1.7 billion years ago.

Etymology and Study

The singular *alga* is the Latin word for a particular seaweed and retains that meaning in English. The etymology is obscure. Although some speculate that it is related to Latin *algçre*, "be cold", there is no known reason to associate seaweed with temperature. A more likely source is *alliga*, "binding, entwining."

The Ancient Greek word for seaweed was *öœêïò* (fûkos or phykos), which could mean either the seaweed (probably red algae) or a red dye derived from it. The Latinization, *fûcus*, meant primarily the cosmetic rouge. The etymology is uncertain, but a strong candidate has long been some word related to the Biblical *ôåê* (pûk), "paint" (if not that word itself), a cosmetic eye-shadow used by the ancient Egyptians and other inhabitants of the eastern Mediterranean. It could be any color: black, red, green, blue.

Accordingly the modern study of marine and freshwater algae is called either phycology or algology, depending on whether the Greek or Latin root is used. The name *Fucus* appears in a number of taxa.

Relationship to Higher Plants

The first plants on earth probably evolved from shallow freshwater algae much like *Chara* some 400 million years ago. These probably had an isomorphic alternation of generations and were probably filamentous. Fossils of isolated land plant spores suggest land plants may have been around as long as 475 million years ago.

Morphology

A range of algal morphologies are exhibited, and convergence of features in unrelated groups is common. The only groups to exhibit three dimensional multicellular thalli are the reds and browns, and some chlorophytes.

Apical growth is constrained to subsets of these groups: the florideophyte reds, various browns, and the charophytes.

The form of charophytes is quite different to those of reds and browns, because have distinct nodes, separated by internode 'stems'; whorls of branches reminiscent of the horsetails occur at the nodes.

Conceptacles are another polyphyletic trait; they appear in the coralline algae and the Hildenbrandiales, as well as the browns.

Most of the simpler algae are unicellular flagellates or amoeboids, but colonial and non-motile forms have developed independently among several of the groups. Some of the more common organizational levels, more than one of which may occur in the life cycle of a species, are

- *Colonial*: small, regular groups of motile cells
- *Capsoid*: individual non-motile cells embedded in mucilage
- *Coccoid*: individual non-motile cells with cell walls
- *Palmelloid*: non-motile cells embedded in mucilage
- *Filamentous*: a string of non-motile cells connected together, sometimes branching
- *Parenchymatous*: cells forming a thallus with partial differentiation of tissues.

In three lines even higher levels of organization have been reached, with full tissue differentiation. These are the brown algae,—some of which may reach 50 m in length (kelps)—the red algae, and the green algae.

The most complex forms are found among the green algae, in a lineage that eventually led to the higher land plants. The point where these non-algal plants begin and algae stop is usually taken to be the presence of reproductive organs with protective cell layers, a characteristic not found in the other alga groups.

Physiology

Many algae, particularly members of the characeae, have served as model experimental organisms to understand the mechanisms of the water permeability of membranes, osmoregulation, turgor regulation, salt tolerance, cytoplasmic streaming, and the generation of action potentials.

Symbiotic Algae

Some species of algae form symbiotic relationships with other organisms. In these symbioses, the algae supply photosynthates (organic substances) to the host organism providing protection to the algal cells. The host organism derives some or all of its energy requirements from the algae. Examples are as follows.

Lichens

Lichens : Composite organisms consisting of a fungus (the mycobiont) and a photosynthetic partner (the photobiont or phycobiont) growing together in a symbiotic relationship. The photobiont is usually either a green alga (commonly *Trebouxia*) or cyanobacterium (commonly *Nostoc*). The morphology, physiology and biochemistry of lichens are very different from those of the isolated fungus and alga in culture. Lichens occur in some of the most extreme environments on Earth—arctic tundra, hot deserts, rocky coasts, and toxic slag heaps. However, they are also abundant as epiphytes on leaves and branches in rain forests and temperate woodland, on bare rock, including walls and gravestones, and on exposed soil surfaces (e.g., *Collema*) in otherwise mesic habitats. The roofs of many buildings have lichens growing on them. Lichens are widespread and may be long-lived; however, many are also vulnerable to environmental disturbance, and may be useful to scientists in assessing the effects of air pollution, ozone depletion, and metal contamination. Lichens have also been used in making dyes and perfumes, as well as in traditional medicines.

Lichens are defined by the International Association for Lichenology to be "an association of a fungus and a photosynthetic symbiont resulting in a stable vegetative body having a specific structure." The fungi, or mycobionts, are from the Ascomycota with a few from the Basidiomycota. They are not found alone in nature but when they began to associate is not known. One mycobiont associates with the same phycobiont species, rarely two, from the green algae, except that alternatively the mycobiont may associate with the same species of cyanobacteria (hence "photobiont" is the more accurate term). A photobiont may be associated with many specific mycobionts or live independently; accordingly, lichens are named and classified as fungal species. The association is termed a morphogenesis because the lichen has a form and capabilities not possessed by the symbiont species alone (they can be experimentally isolated). It is possible that the photobiont triggers otherwise latent genes in the mycobiont.

Overview

The body (thallus) of most lichens is different from those of either the fungus or alga growing separately. The fungus surrounds the algal cells, often enclosing them within complex fungal tissues unique to lichen associations. In many species the fungus penetrates the algal cell wall, forming penetration pegs or haustoria similar to those produced by pathogenic fungi. Lichens are poikilohydric, capable of surviving extremely low levels of water content. However, the re-configuration of membranes following a period of dehydration requires several minutes at least.

The algal or cyanobacterial cells are photosynthetic, and as in plants they reduce atmospheric carbon dioxide into organic carbon sugars to feed both symbionts. Both partners gain water and mineral nutrients mainly from the atmosphere, through rain and dust. The fungal partner protects the alga by retaining water, serving as a larger capture area for mineral nutrients and, in some cases, provides minerals obtained from the substrate. If a cyanobacterium is present, as a primary partner or another symbiont in addition to green alga as in certain tripartite lichens, they can fix atmospheric nitrogen, complementing the activities of the green alga.

Although strains of cyanobacteria found in various cyanolichens are often closely related to one another, they differ from the most closely related free-living strains. The lichen association is a close symbiosis. It extends the ecological range of both partners but is not always obligatory for their growth and reproduction in natural environments, since many of the algal symbionts can live independently.

A prominent example is the alga *Trentepohlia* which forms orange-coloured populations on tree trunks and suitable rock faces. Lichen propagules (diaspores) typically contain cells from both partners, although the fungal components of so-called "fringe species" rely instead on algal cells dispersed by the "core species." Lichen associations may be examples of mutualism, commensalism or even parasitism, depending on the species. Cyanobacteria in laboratory settings can grow faster when they are alone rather than when they are part of a lichen. Recently, on 26 April 2012, scientists reported that lichen survived and showed remarkable results on the adaptation capacity of photosynthetic activity within the simulation time of 34 days under Martian conditions in the Mars Simulation Laboratory (MSL) maintained by the German Aerospace Centre (DLR).

History

Although lichens had been recognized as organisms for quite some time, it was not until 1867, when Swiss botanist Simon Schwendener proposed his dual theory of lichens, that the true nature of the lichen association began to emerge. Schwendener's hypothesis, which at the time lacked experimental evidence, arose from his extensive analysis of the anatomy and development in lichens, algae, and fungi using a light microscope. Many of the leading lichenologists at the time, such as James Crombie and Nylander, rejected Schwendener's hypothesis because the common consensus was that all living organisms were autonomous. Other prominent biologists, such as Heinrich Anton de Bary, Albert Bernhard Frank, Melchior Treub and Hermann Hellriegel were not so quick to reject Schwendener's ideas and the concept soon spread into other areas of study, such as microbial, plant, animal and human pathogens. When the complex relationships between pathogenic microorganisms and their hosts were finally identified—refuting the idea of holistic organisms—Schwendener's hypothesis began to gain popularity. Further experimental proof of the dual nature of lichens was obtained when Eugen Thomas published his results in 1939 on the first successful re-synthesis experiment.

Symbionts

Living as a symbiont in a lichen appears to be a very successful way for a fungus to derive essential nutrients, as about 20% of all fungal species have acquired this mode of life. The largest number of lichenized fungi occur in the Ascomycota, with about 40% of species forming such an association. Some of these lichenized fungi occur in orders with nonlichenized fungi that live as saprotrophs or plant parasites (for example, the Leotiales, Dothideales, and Pezizales). Other lichen fungi occur in only five orders in which all members are engaged in this habit (Orders Graphidales, Gyalectales, Peltigerales, Pertusariales, and Teloschistales). Lichenized and nonlichenized fungi can even be found in the same genus or species. Overall, about 98% of lichens have an ascomycetous mycobiont . Next to the Ascomycota, the largest number of lichenized fungi occur in the unassigned fungi imperfecti. Comparatively few Basidiomycetes are lichenized, but these include agarics, such as species of *Lichenomphalia*, clavarioid fungi, such as species of *Multiclavula*, and corticioid fungi, such as species of *Dictyonema*.

The autotrophic symbionts occurring in lichens are simple, photosynthetic organisms commonly and traditionally known as algae. These symbionts include both prokaryotic and eukaryotic organisms. Approximately 100 species of photosynthetic partners from 40 genera and five distinct classes (prokaryotic: Cyanophyceae; eukaryotic: Trebouxiophyceae, Phaeophyceae, Chlorophyceae) have been found to associate with the lichen-forming fungi. The prokaryotes belong to the Cyanobacteria, whose representatives are often called bluegreen algae. The bluegreen algae occur as symbionts in about 8% of the known lichens. The most commonly occurring genus is *Nostoc*. The majority of the lichens contain eukaryotic autotrophs belonging to the Chlorophyta (green algae) or to the Xanthophyta (yellow-green algae). About 90% of all known lichens have a green alga as a symbiont, and among these, *Trebouxia* is the most common genus, occurring in about 40% of all lichens. The second most commonly represented green alga genus is *Trentepohlia*. Overall, about 100 species are known to occur as autotrophs in lichens. All the algae are probably able to exist independently in nature as well as in the lichen.

A particular fungus species and algal species are not necessarily always associated together in a lichen. One fungus, for example, can form lichens with a variety of different algae. The thalli produced by a given fungal symbiont with its differing partners will be similar, and the secondary metabolites identical, indicating that the fungus has the dominant role in determining the morphology of the lichen. Further, the same algal species can occur in association with different fungal partners. Lichens are known in which there is one fungus associated with two or even three algal species. Rarely, the reverse can occur, and two or more fungal species can interact to form the same lichen.

Both the lichen and the fungus partner bear the same scientific name, and the lichens are being integrated into the classification schemes for fungi. The alga bears its own scientific name, which bears no relationship to that of the lichen or fungi.

Morphology and Structure

Some lichens have the aspect of leaves (foliose lichens); others cover the substrate like a crust (crustose lichens) (*illustration, right*), others such as the genus *Ramalina* adopt shrubby forms (fruticose lichens), and there are gelatinous lichens such as the genus *Collema*.

Although the form of a lichen is determined by the genetic material of the fungal partner, association with a photobiont is required for the development of that form. When grown in the laboratory in the absence of its photobiont, a lichen fungus develops as an undifferentiated mass of hyphae. If combined with its photobiont under appropriate conditions, its characteristic form emerges, in the process called morphogenesis (Brodo, Sharnoff & Sharnoff, 2001). In a few remarkable cases, a single lichen fungus can develop into two very different lichen forms when associating with either a green algal or a cyanobacterial symbiont. Quite naturally, these alternative forms were at first considered to be different species, until they were first found growing in a conjoined manner.

There is evidence to suggest that the lichen symbiosis is parasitic or commensalistic, rather than mutualistic (Ahmadjian 1993). The photosynthetic partner can exist in nature independently of the fungal partner, but not vice versa. Furthermore, photobiont cells are routinely destroyed in the course of nutrient exchange. The association is able to continue because reproduction of the photobiont cells matches the rate at which they are destroyed. (ibid.) Under magnification, a section through a typical foliose lichen thallus reveals four layers of interlaced fungal filaments. The uppermost layer is formed by densely agglutinated fungal hyphae building a protective outer layer called the cortex, which can reach several hundred ìm in thickness. This cortex may be further topped by an epicortex 0.6-1ìm thick in some Parmeliaceae, which may be with or without pores, and is secreted by cells—it is not itself cellular. In lichens that include both green algal and cyanobacterial symbionts, the cyanobacteria may be held on the upper or lower surface in small pustules called cephalodia. Beneath the upper cortex is an algal layer composed of algal cells embedded in rather densely interwoven fungal hyphae. Each cell or group of cells of the photobiont is usually individually wrapped by hyphae, and in some cases penetrated by an haustorium. Beneath this algal layer is a third layer of loosely interwoven fungal hyphae without algal cells.

This layer is called the medulla. Beneath the medulla, the bottom surface resembles the upper surface and is called the lower cortex, again consisting of densely packed fungal hyphae. The lower cortex often bears rootlike fungal structures known as rhizines, which serve to attach the thallus to the substrate on which it grows.

Lichens also sometimes contain structures made from fungal metabolites, for example crustose lichens sometimes have a polysaccharide layer in the cortex. Although each lichen thallus generally appears homogeneous, some evidence seems to suggest that the fungal component may consist of more than one genetic individual of that species. This seems to also be true of the photobiont species involved.

Growth Form

Lichens are informally classified by growth form into:

- crustose (paint-like, flat), e.g., *Caloplaca flavescens*
- filamentous (hair-like), e.g., *Ephebe lanata*
- foliose (leafy), e.g., *Hypogymnia physodes*
- fruticose (branched), e.g., *Cladonia evansii, C. subtenuis,* and *Usnea australis*
- leprose (powdery), e.g., *Lepraria incana*
- squamulose (consisting of small scale-like structures, lacking a lower cortex), e.g., *Normandina pulchella*
- gelatinous lichens, in which the cyanobacteria produce a polysaccharide that absorbs and retains water.

Reproduction and Dispersal

Many lichens reproduce asexually, either by vegetative reproduction or through the dispersal of diaspores containing algal and fungal cells. *Soredia* (singular soredium) are small groups of algal cells surrounded by fungal filaments that form in structures called *soralia*, from which the soredia can be dispersed by wind. Another form of diaspore are *isidia*, elongated outgrowths from the thallus that break off for mechanical dispersal. Fruticose lichens in particular can easily fragment. Because of the relative lack of differentiation in the thallus, the line between diaspore formation and vegetative reproduction is often blurred. Many lichens break up into fragments when they dry, dispersing themselves by wind action, to resume growth when moisture returns.

Many lichen fungi appear to reproduce sexually in a manner typical of fungi, producing spores that are presumably the result of sexual fusion and meiosis.

Following dispersal, such fungal spores must meet with a compatible algal partner before a functional lichen can form. This may be a common form of reproduction in basidiolichens, which form fruitbodies resembling their nonlichenized relatives. Among the ascolichens, spores are produced in spore-producing bodies, the three most common spore body types are the apothecia, perithecia and the pycnidia.

For reproduction, lichen possess isidia, soredia, and undergo simple fragmentation. These structures are also composed of a fungal hyphae wrapped around cyanobacteria. (Eichorn, Evert, and Raven, 2005) While the reproductive structures are all composed of the same components (Mycobiont and Photobiont) they are each unique in other ways. Isidia are small outgrowths on the exterior of the lichen. Soredia are powdery propagules that are released from the top of the thallus. In order to establish the lichen, the soredia propagules must contain both the photobiont and the mycobiont.

Growth and longevity

Lichenometry

Lichenometry is a technique used to determine the age of exposed rock surfaces based on the size of lichen thalli.

Introduced by Beschel in the 1950s, the technique has found many applications.

Ecology

Lichens must compete with plants for access to sunlight, but because of their small size and slow growth, they thrive in places where higher plants have difficulty growing. Lichens are often the first to settle in places lacking soil, constituting the sole vegetation in some extreme environments such as those found at high mountain elevations and at high latitudes. Some survive in the tough conditions of deserts, and others on frozen soil of the Arctic regions.

A major ecophysiological advantage of lichens is that they are poikilohydric (*poikilo-* variable, *hydric-* relating to water), meaning that though they have little control over the status of their hydration, they can tolerate irregular and extended periods of severe desiccation. Like some mosses, liverworts, ferns, and a few "resurrection plants", upon desiccation, lichens enter a metabolic suspension or stasis (known as cryptobiosis) in which the cells of the lichen symbionts are dehydrated to a degree that halts most biochemical activity. In this cryptobiotic state, lichens can survive wider extremes of temperature, radiation and drought in the harsh environments they often inhabit.

Lichens do not have roots and do not need to tap continuous reservoirs of water like most higher plants, thus they can grow in locations impossible for most plants, such as bare rock, sterile soil or sand, and various artificial structures such as walls, roofs and monuments. Many lichens also grow as epiphytes (*epi-* on the surface, *phyte-* plant) on plants, particularly on the trunks and branches of trees. When growing on plants, lichens are not parasites; they do not consume any part of the plant nor poison it.

Some ground-dwelling lichens, such as members of the subgenus *Cladina* (reindeer lichens), however, produce allelopathic chemicals which leach into the soil and inhibit the germination of plant seeds and growth of young plants. Stability (that is, longevity) of their substrate is a major factor of lichen habitats. Most lichens grow on stable rock surfaces or the bark of old trees, but many others grow on soil and sand. In these latter cases, lichens are often an important part of soil stabilization; indeed, in some desert ecosystems, vascular (higher) plant seeds cannot become established except in places where lichen crusts stabilize the sand and help retain water.

The European Space Agency has discovered that lichens can survive unprotected in space. In an experiment led by Leopoldo Sancho from the Complutense University of Madrid, two species of lichen—*Rhizocarpon geographicum* and *Xanthoria elegans*—were sealed in a capsule and launched on a Russian Soyuz rocket on 31 May 2005. Once in orbit the capsules were opened and the lichens were directly exposed to the vacuum of space with its widely fluctuating temperatures and cosmic radiation. After 15 days the lichens were brought back to earth and were found to be in full health with no discernible damage from their time in orbit.

When growing on mineral surfaces, some lichens slowly decompose their substrate by chemically degrading and physically disrupting the minerals, contributing to the process of weathering by which rocks are gradually turned into soil. While this contribution to weathering is usually benign, it can cause problems for artificial stone structures. For example, there is an ongoing lichen growth problem on Mount Rushmore National Memorial that requires the employment of mountain-climbing conservators to clean the monument.

Lichens may be eaten by some animals, such as reindeer, living in arctic regions. The larvae of a number of Lepidoptera species feed exclusively on lichens. These include Common Footman and Marbled Beauty. However, lichens are very low in protein and high in carbohydrates, making them unsuitable for some animals. Lichens are also used by the Northern Flying Squirrel for nesting, food, and a water source during winter.

Air Pollution

If lichens are exposed to air pollutants at all times, without any deciduous parts, they are unable to avoid the accumulation of pollutants. Also lacking stomata and a cuticle, lichens may absorb aerosols and gases over the entire thallus surface from which they may readily diffuse to the photobiont layer. Because lichens do not possess roots, their primary source of most elements is the air, and therefore elemental levels in lichens often reflect the accumulated composition of ambient air. The processes by which atmospheric deposition occurs include fog and dew, gaseous absorption, and dry deposition. Consequently, many environmental studies with lichens emphasize their feasibility as effective biomonitors of atmospheric quality.

Not all lichens are equally sensitive to air pollutants, so different lichen species show different levels of sensitivity to specific atmospheric pollutants.

The sensitivity of a lichen to air pollution is directly related to the energy needs of the mycobiont, so that the stronger the dependency of the mycobiont on the photobiont, the more sensitive the lichen is to air pollution. Upon exposure to air pollution, the photobiont may use metabolic energy for repair of cellular structures that would otherwise be used for maintenance of photosynthetic activity, therefore leaving less metabolic energy available for the mycobiont. The alteration of the balance between the photobiont and mycobiont can lead to the breakdown of the symbiotic association. Therefore, lichen decline may result not only from the accumulation of toxic substances, but also from altered nutrient supplies that favour one symbiont over the other.

Evolution and Paleontology

The evolution of lichens and the phylum Ascomycota is complex and not well understood, but because there are fifteen different classes of Ascomycetes, scientists generally believe that different lichens have evolved independently from one another through analogous evolution. Lichenized fungi have continued to evolve, developing differently than those that do not form lichens.

Lichenization is an ancient nutritional strategy for fungi. The extreme habitats that lichens inhabit are not ordinarily conducive to producing fossils. The oldest fossil lichens in which both symbiotic partners have been recovered date to the Early Devonian Rhynie chert, about 400 million years old. The slightly older fossil *Spongiophyton* has also been interpreted as a lichen on morphological and isotopic grounds, although the isotopic basis is decidedly shaky. It has been suggested—although not yet proven—that the even older fossil *Nematothallus* was a lichen.

It has also been claimed that Ediacaran fossils were lichens;, though this claim is controversial. Additional evidence has been marshalled for a lichen interpretation of *Dickinsonia*. Lichen-like fossils consisting of coccoid cells and thin filaments, preserved in marine phosphorite of the Doushantuo Formation in southern China. These fossils are thought to be 551 to 635 million years old (belonging to the Neoproterozoic era). Discovery of these fossils suggest that fungi developed symbiotic partnerships with photoautotrophs long before the evolution of vascular plants.

Winfrenatia, an early zygomycetous lichen symbiosis that may have involved controlled parasitism, is an impression found in Scotland, belonging to the early Devonian times. There are also several examples of fossilized lichens embedded in amber. The fossilized *Anzia* is found in pieces of amber in northern Europe and dates back approximately 40 million years. Fossilized *Lobaria* comes from Trinity County in northern California, USA and dates back to the early to middle Miocene.

In 1995, Gargas and colleagues proposed that there were at least five independent origins of lichenization; three in the basidiomycetes and at least two in the Ascomycetes. However, Lutzoni et al. (2000) indicate that lichenization probably evolved earlier and was followed by multiple independent losses. Some non-lichen-forming fungi may have secondarily lost the ability to form a lichen association. As a result, lichenization has been viewed as a highly successful nutritional strategy.

Lichens were a component of the early terrestrial ecosystems, and the estimated age of the oldest terrestrial lichen fossil is 400 Ma. Recent (2009) studies suggest that the ancestral ecological state of the Ascomycota was saprobism, and that independent lichenization events have occurred multiple times.

Taxonomy and Classification

Lichens are named based on the fungal component, which plays the primary role in determining the lichen's form. The fungus typically comprises the majority of a lichen's bulk, though in filamentous and gelatinous lichens this is not always the case. The lichen fungus is typically a member of the Ascomycota—rarely a member of the Basidiomycota, and then termed basidiolichens to differentiate them from the more common ascolichens. Formerly, some lichen taxonomists placed lichens in their own division, the Mycophycophyta, but this practice is no longer accepted because the components belong to separate lineages. Neither the ascolichens nor the basidiolichens form monophyletic lineages in their respective fungal phyla, but they do form several major solely or primarily lichen-forming groups within each phylum. Even more unusual than basidiolichens is the fungus *Geosiphon pyriforme*, a member of the Glomeromycota that is unique in that it encloses a cyanobacterial symbiont inside its cells. *Geosiphon* is not usually considered to be a lichen, and its peculiar symbiosis was not recognized for many years. The genus is more closely allied to endomycorrhizal genera. The following table lists the orders and families of fungi that include lichen-forming species.

Economic Uses

Food

Lichens are eaten by many different cultures across the world. Although some lichens are only eaten in times of famine, others are a staple food or even a delicacy. Two obstacles are often encountered when eating lichens: lichen polysaccharides are generally indigestible to humans, and lichens usually contain mildly toxic secondary compounds that should be removed before eating. Very few lichens are poisonous, but those high in vulpinic acid or usnic acid are toxic. Most poisonous lichens are yellow.

In the past Iceand moss (*Cetraria islandica*) was an important human food in northern Europe, and was cooked as a bread, porridge, pudding, soup, or salad. Wila (*Bryoria fremontii*) was an important food in parts of North America, where it was usually pitcooked. Northern peoples in North America and Siberia traditionally eat the partially digested reindeer lichen (*Cladina* spp.) after they remove it from the rumen of caribou or reindeer that have been killed. Rock tripe (*Umbilicaria* spp. and *Lasalia* spp.) is a lichen that has frequently been used as an emergency food in North America, and one species, *Umbilicaria esculenta*, is used in a variety of traditional Korean and Japanese foods.

Other Uses

Many lichens produce secondary compounds, including pigments that reduce harmful amounts of sunlight and powerful toxins that reduce herbivory or kill bacteria. These compounds are very useful for lichen identification, and have had economic importance as dyes such as cudbear or primitive antibiotics.

There are reports dating almost 2000 years old of lichens being used to extract purple and red colors. Of great historical and commercial significance are lichens belonging to the family Roccellaceae, commonly called orchella weed or orchil. Orcein and other lichen dyes have largely been replaced by synthetic versions. The pH indicator litmus is a dye extracted from the lichen genus *Rocella tinctoria* by boiling

Extracts from many *Usnea* species were used to treat wounds in Russia in the mid-twentieth century. Historically in Europe, *Lobaria pulmonaria* was collected in large quantities as "Lungwort" and, due to its lung-like appearance, was sold as a cure for lung diseases. Similarly *Peltigera leucophlebia* was used as a supposed cure for thrush, due to the resemblance of its cephalodia to the appearance of the disease. The substance olivetol is found to be naturally present in certain species of lichens. This is a property it shares with the cannabis plant, which internally produces the related substance olivetolic acid (before using it to biosynthesise tetrahydrocannabinol (THC)). Lichen is used in model railroading and other modelling hobbies as a material for making trees and shrubs.

Coral Reefs

Coral reefs are accumulated from the calcareous exoskeletons of marine invertebrates of the order Scleractinia (stony corals). As animals they metabolize sugar and oxygen to obtain energy for their cell-building processes, including secretion of the exoskeleton, with water and carbon dioxide as byproducts. As the reef is the result of a favourable equilibrium between construction by the corals and destruction by marine erosion, the rate at which metabolism can proceed determines the growth or deterioration of the reef. Dinoflagellates (algal protists) are often endosymbionts in the cells of marine invertebrates, where they accelerate host-cell metabolism by generating immediately available sugar and oxygen through photosynthesis using incident light and the carbon dioxide produced by the host. Stony corals that are reef-building corals (hermatypic corals) require endosymbiotic algae from the genus *Symbiodinium* to be in a healthy condition. The loss of *Symbiodinium* from the host is known as coral bleaching, a condition which leads to the deterioration of a reef.

Sea Sponges

Green algae live close to the surface of some sponges, for example, breadcrumb sponge (*Halichondria panicea*). The alga is thus protected from predators; the sponge is provided with oxygen and sugars which can account for 50 to 80% of sponge growth in some species.

Life-cycle

Rhodophyta, Chlorophyta and Heterokontophyta, the three main algal Phyla, have life-cycles which show tremendous variation with considerable complexity. In general there is an asexual phase where the seaweed's cells are diploid, a sexual phase where the cells are haploid followed by fusion of the male and female gametes. Asexual reproduction is advantageous in that it permits efficient population increases, but less variation is possible. Sexual reproduction allows more variation, but is more costly. Often there is no strict alternation between the sporophyte and also because there is often an asexual phase, which could include the fragmentation of the thallus.

Conceptacle

Conceptacles are specialised cavities of marine and freshwater algae that contain the reproductive organs. They are situated in the receptacle and open by a small ostiole. Conceptacles are present in Corallinaceae, and Hildenbrandiales, as well as the brown Fucales. In the Fucales there is no haploid phase in the reproductive cycle and therefore no alternation of generations. The thallus is a sporophyte. The diploid plants produce male (antheridia) and female (oogonia) gametangia by meiosis. The gametes are released into the sea, after fusion the zygote settles and begins growth. Two taxa, the corallines and Hildenbrandiaceae, bear conceptacles, although the striking difference between their formation indicates that the conceptacles are not homologous. Similar structures also exist: *cryptostomata* are similar to conceptacles but differ having only hairs and are sterile. *Caecostomata*, are found only in *Fucus distichus*, in these the ostiole becomes blocked during development.

Morphology

Conceptacles can measure from 40–150 ìm in diameter in some Hildenbrandiales. Some conceptacles are pits within the thallus, whereas others form raised domes protruding above it. Conceptacles open to the environment via one or many pores, depending on the species and the type of conceptacle.

Components

Columella: This is a central pillar constituting central calcified cells covered by cellular debris. These sterile cells rise up from the conceptacle floor to form a peak.

Roof Filaments: Some conceptacles form by the centripetal expansion of a hole near the thallus surface; in such cases, a roof forms by nearby filaments arching over and establishing themselves as short (often 1–9 cells long) filaments that cover the chamber, leaving a central pore through which the spores can escape. If these filaments are vicarious they may produce a beak-like opening.

Spores: In asexual conceptacles, the spores tend to be large, and so are squeezed in, filling the conceptacle chamber "like orange segments".

Types of Conceptacle

There are a range of different conceptacles, classified according to the nature of the spores that they contain; some species may possess as many as four distinct types. For example, *Lithophyllum incrustans* bears asexual, male, female and cystocarpic (strictly, a type of female) conceptacles, discussed below.

Asexual Conceptacles: These develop where weakly calcified cells break down to produce a cavity. In *L. incrustans*, Their columella is central and obvious; it is accentuated by the spores (which may be bispores, tetraspores, etc.), which squeeze against the columella and conceptacle walls. Asexual conceptacles may be uniporate or multiporate.

Sexual (Male / Female) Conceptacles: These lack a columella. *L. incrustans* has distinct male and female plants; the two conceptacle types never co-occur on the same thallus. Male conceptacles tend not to sit as deep within the thallus as female conceptacles; their size is similar although their shape tends to differ. Cells within circular regions within the perithallus lengthen to form sexual conceptacles, leaving long, wefty cells around the walls of the conceptacle. In the male conceptacles of *Austrolithon*, marginal filaments grow more quickly, bending in over the top of the conceptacle cavity and eventually setting up as distinct filaments that form the conceptacle roof.

Cystocarpic Conceptacles: These are not associated with tissue demineralization; rather, they start to form at the centre (as a female conceptacle) and develop radially. Being mature female conceptacles, these too lack a columella.

Development

In most coralline algae, a cluster of reproductive cells forms in the middle layer of the alga, and is engulfed by the surrounding tissue, which grows up and over the reproductive cells to form a roof and a uniporate conceptacle. Caps may subsequently develop to protect the opening. However the conceptacle may originate at any depth within the thallus, at the surface layer or at the basal perithallus.

Four different modes of asexual conceptacle formation exist. In *Corallina* and *Bossiella*. In the coralline *Bossiella*, the conceptacle instead forms in the outer layer.

A thickening forms, which separates the outer epithallium from the underlying cortex; this thickening and the overlying epithallium will end up being the cap of the conceptacle, and the underlying cells will develop to become reproductive initials. Once the cap is formed, the filamentous cells underneath begin to degrade. The tissue around the edge of the cap grows more quickly. This combination produces a chamber beneath the cap.

Evolutionary History

Conceptacles appear in the fossil record at least as early as the Silurian. Similar structures have been noted in *Prototaxites*, which would imply that this giant land organism was not a simple fungus (as most paleontologists assume today) but a lichen.

Numbers

The *Algal Collection of the US National Herbarium* (located in the National Museum of Natural History) consists of approximately 320,500 dried specimens, which, although not exhaustive (no exhaustive collection exists), gives an idea of the order of magnitude of the number of algal species (that number remains unknown). Estimates vary widely. For example, according to one standard textbook, in the British Isles the *UK Biodiversity Steering Group Report* estimated there to be 20000 algal species in the UK. Another checklist reports only about 5000 species. Regarding the difference of about 15000 species, the text concludes: "It will require many detailed field surveys before it is possible to provide a reliable estimate of the total number of species ..."

Regional and group estimates have been made as well:

- 5000–5500 species of red algae worldwide
- "some 1300 in Australian Seas"
- 400 seaweed species for the western coastline of South Africa, and 212 species from the coast of KwaZulu-Natal. Some of these are duplicates as the range extends across both coasts, and the total recorded is probably about 500 species. Most of these are listed in List of seaweeds of South Africa. These exclude phytoplankton and crustose corallines.
- 669 marine species from California (US)
- 642 in the check-list of Britain and Ireland and so on, but lacking any scientific basis or reliable sources, these numbers have no more credibility than the British ones mentioned above. Most estimates also omit microscopic algae, such as phytoplankton. The most recent estimate suggests a total number of 72,500 algal species worldwide.

Distribution

The topic of distribution of algal species has been fairly well studied since the founding of phytogeography in the mid-

19th century AD. Algae spread mainly by the dispersal of spores analogously to the dispersal of Plantae by seeds and spores. Spores are everywhere in all parts of the Earth: the waters fresh and marine, the atmosphere, free-floating and in precipitation or mixed with dust, the humus and in other organisms, such as humans. Whether a spore is to grow into an organism depends on the combination of the species and the environmental conditions of where the spore lands.

The spores of fresh-water algae are dispersed mainly by running water and wind, as well as by living carriers. The bodies of water into which they are transported are chemically selective. Marine spores are spread by currents. Ocean water is temperature selective, resulting in phytogeographic zones, regions and provinces.

To some degree the distribution of algae is subject to floristic discontinuities caused by geographical features, such as Antarctica, long distances of ocean or general land masses. It is therefore possible to identify species occurring by locality, such as "Pacific Algae" or "North Sea Algae". When they occur out of their localities, it is usually possible to hypothesize a transport mechanism, such as the hulls of ships. For example, *Ulva reticulata* and *Ulva fasciata* travelled from the mainland to Hawaii in this manner.

Mapping is possible for select species only: "there are many valid examples of confined distribution patterns." For example, *Clathromorphum* is an arctic genus and is not mapped far south of there. On the other hand, scientists regard the overall data as insufficient due to the "difficulties of undertaking such studies."

Locations

Algae are prominent in bodies of water, common in terrestrial environments and are found in unusual environments, such as on snow and on ice. Seaweeds grow mostly in shallow marine waters, under 100 metres (330 ft); however some have been recorded to a depth of 360 metres (1,180 ft). The various sorts of algae play significant roles in aquatic ecology. Microscopic forms that live suspended in the water column (phytoplankton) provide the food base for most marine food chains. In very high densities (algal blooms) these algae may discolor the water and outcompete, poison, or asphyxiate other life forms.

Algae are variously sensitive to different factors, which has made them useful as biological indicators in the Ballantine Scale and its modification.

Uses

Agar

Agar, a gelatinous substance derived from red algae, has a number of commercial uses. It is a good medium for bacteria.

Alginates

Between 100,000 and 170,000 wet tons of *Macrocystis* are harvested annually in California for alginate extraction and abalone feed.

Energy Source

To be competitive and independent from fluctuating support from (local) policy on the long run, biofuels should equal or beat the cost level of fossil fuels. Here, algae based fuels hold great promise, directly related to the potential to produce more biomass per unit area in a year than any other form of biomass. The break-even point for algae-based biofuels is estimated to occur in about ten to fifteen years.

Fertilizer

For centuries seaweed has been used as a fertilizer; George Owen of Henllys writing in the 16th century referring to drift weed in South Wales: This kind of ore they often gather and lay on great heapes, where it heteth and rotteth, and will have a strong and loathsome smell; when being so rotten they cast on the land, as they do their muck, and thereof springeth good corn, especially barley ... After spring-tydes or great rigs of the sea, they fetch it in sacks on horse backes, and carie the same three, four, or five miles, and cast it on the lande, which doth very much better the ground for corn and grass.

Today, algae are used by humans in many ways; for example, as fertilizers, soil conditioners and livestock feed. Aquatic and microscopic species are cultured in clear tanks or ponds and are either harvested or used to treat effluents pumped through the ponds. Algaculture on a large scale is an important type of aquaculture in some places. Maerl is commonly used as a soil conditioner.

Nutrition

Naturally growing seaweeds are an important source of food, especially in Asia. They provide many vitamins including: A, B_1, B_2, B_6, niacin and C, and are rich in iodine, potassium, iron, magnesium and calcium. In addition commercially cultivated microalgae, including both algae and cyanobacteria, are marketed as nutritional supplements, such as Spirulina, Chlorella and the Vitamin-C supplement, Dunaliella, high in beta-carotene.

Algae are national foods of many nations: China consumes more than 70 species, including *fat choy*, a cyanobacterium considered a vegetable; Japan, over 20 species; Ireland, dulse; Chile, cochayuyo. Laver is used to make "laver bread" in Wales where it is known as *bara lawr*; in Korea, gim; in Japan, nori and aonori. It is also used along the west coast of North America from California to British Columbia, in Hawaii and by the Mâori of New Zealand. Sea lettuce and badderlocks are a salad ingredient in Scotland, Ireland, Greenland and Iceland.

The oils from some algae have high levels of unsaturated fatty acids. For example, *Parietochloris incisa* is very high in arachidonic acid, where it reaches up to 47% of the triglyceride pool. Some varieties of algae favoured by vegetarianism and veganism contain the long-chain, essential omega-3 fatty acids, Docosahexaenoic acid (DHA) and Eicosapentaenoic acid (EPA). Fish oil contains the omega-3 fatty acids, but the original source is algae (microalgae in particular), which are eaten by marine life such as copepods and are passed up the food chain. Algae has emerged in recent years as a popular source of omega-3 fatty acids for vegetarians who cannot get long-chain EPA and DHA from other vegetarian sources such as flaxseed oil, which only contains the short-chain Alpha-Linolenic acid (ALA).

Edible Seaweed

Edible seaweed are algae that can be eaten and used in the preparation of food. It typically contains high amounts of fibre and, contrary to land based plant foods, they contain a complete protein. They may belong to one of several groups of multicellular algae: the red algae, green algae, and brown algae.

Seaweeds are also harvested or cultivated for the extraction of alginate, agar and carrageenan, gelatinous substances collectively known as hydrocolloids or phycocolloids. Hydrocolloids have attained commercial significance, especially in food production as food additives. The food industry exploits the gelling, water-retention, emulsifying and other physical properties of these hydrocolloids.

Most edible seaweeds are marine algae as most freshwater algae are toxic. While marine algae are not toxic, some do contain acids that irritate the digestion canal, while some others can have a laxative and electrolyte-balancing effect.

Distribution

Seaweeds are used extensively as food in coastal cuisines around the world. Seaweed has been a part of diets in China, Japan, and Korea since prehistoric times. Seaweed is also consumed in many traditional European societies, in Iceland and western Norway, the Atlantic coast of France, northern and western Ireland, Wales and some coastal parts of South West England, as well as Nova Scotia and Newfoundland. The Mâori people of New Zealand traditionally used a few species of red and green seaweed.

Nutrition and Uses

Seaweed contains high levels of iodine relative to other foods. In the Philippines, Tiwi, Albay residents discovered a new pancit or noodles made from seaweed, which can be cooked into pancit canton, pancit luglug, spaghetti or carbonara and is claimed to have health benefits such as being rich in calcium, magnesium and iodine.

Polysaccharides in seaweed may be metabolized in humans through the action of bacterial gut enzymes. Research has failed to find such enzymes in the North-American population, while being frequent in the Japanese population.

In some parts of Asia, nori wmÔ, (in Japan), zicai +}Ü*f* (in China), and gim @® (in Korea), sheets of the dried red alga

Porphyra are used in soups or to wrap sushi or onigiri. *Chondrus crispus* (commonly known as Irish moss) is another red alga used in producing various food additives, along with Kappaphycus and various gigartinoid seaweeds.

Japanese cuisine has seven types of seaweed identified by name, and thus the term for seaweed in Japanese is used primarily in scientific applications, and not in reference to food.

Common Edible Seaweeds

Common edible seaweeds include:

- Arame (*Eisenia bicyclis*)
- Badderlocks (*Alaria esculenta*)
- Bladderwrack (*Fucus vesiculosus*)
- Carola (various species of *Callophyllis*)
- Carrageen moss (*Mastocarpus stellatus*)
- Channelled wrack (*Pelvetia canaliculata*)
- Chlorella
- Cochayuyo (*Durvillaea antarctica*)
- Dulse (*Palmaria palmata*)
- Eucheuma spinosum, Euchema cottonii
- Gutweed (*Enteromorpha intestinalis*)
- Gelidiella acerosa
- Gracilaria edulis, Gracilaria corticata
- Hijiki or Hiziki (*Sargassum fusiforme*)
- Hypnea
- Irish moss (*Chondrus crispus*)
- Laver (*Porphyra laciniata/Porphyra umbilicalis*)
- Limu Kala (*Sargassum echinocarpum*)
- Kombu (*Saccharina japonica*)
- Mozuku (*Cladosiphon okamuranus*)
- Nori (various species of the red alga *Porphyra*)
- Oarweed (*Laminaria digitata*)
- Ogonori (*Gracilaria*)
- Sugar kelp (*Saccharina latissima*)

- Sea Grapes or *green caviar* (*Caulerpa lentillifera*)
- Sargassum cinetum, Sargassum vulgare, Sargassum swartzii, Sargassum myriocysum
- Sea Lettuce (various species of the genus *Ulva*)
- Spiral wrack (*Fucus spiralis*)
- Spirulina (*Arthrospira platensis* and *Arthrospira maxima*)
- Thongweed (*Himanthalia elongata*)
- Wakame (*Undaria pinnatifida*)
    - * *Hirome* or *hitohame wakame* (*Undaria undarioides*).

Pollution Control

- Sewage can be treated with algae, reducing the need for greater amounts of toxic chemicals than are already used.
- Algae can be used to capture fertilizers in runoff from farms. When subsequently harvested, the enriched algae itself can be used as fertilizer.
- Aquariums and ponds can be filtered using algae, which absorb nutrients from the water in a device called an algae scrubber, also known as an "ATS".

Agricultural Research Service scientists found that 60-90% of nitrogen runoff and 70-100% of phosphorus runoff can be captured from manure effluents using an algal turf scrubber (ATS). Scientists developed the ATS, which are shallow, 100-foot raceways of nylon netting where algae colonies can form, and studied its efficacy for three years. They found that algae can readily be used to reduce the nutrient runoff from agricultural fields and increase the quality of water flowing into rivers, streams, and oceans. The enriched algae itself also can be used as a fertilizer. Researchers collected and dried the nutrient-rich algae from the ATS and studied its potential as an organic fertilizer. They found that cucumber and corn seedlings grew just as well using ATS organic fertilizer as they did with commercial fertilizers.

Pigments

The natural pigments produced by algae can be used as an alternative to chemical dyes and coloring agents.

Stabilizing Substances

Carrageenan, from the red alga *Chondrus crispus*, is used as a stabilizer in milk products.

Carrageenan

Carrageenans or carrageenins are a family of linear sulfated polysaccharides that are extracted from red seaweeds. They are widely used in the food industry, for their gelling, thickening and stabilizing properties.

Their main application is in dairy and meat products, due to their strong interactions with protein. There are three main varieties of carrageenan, which differ in their degree of sulfation. Kappa-carrageenan has one sulfate per disaccharide. Iota-carrageenan has two sulfates per disaccharide.

Lambda carrageenan has three sulfates per disaccharide. Gelatinous extracts of the *Chondrus crispus* (Irish Moss) seaweed have been used as food additives for hundreds of years.

Carrageenan is a vegetarian and vegan alternative to gelatin in some applications, although it cannot replace gelatin in confectionery like jelly babies. Scientists have raised serious concerns about the safety of carrageenan in food, based on laboratory animal studies showing gastrointestinal inflammation, ulcerations and colitis-like disease in animals given food-grade carrageenan in their drinking water or diet. Some physicians advise avoiding consumption of foods with carrageenan, especially for people with gastrointestinal symptoms..

Properties

Carrageenans are large, highly flexible molecules that curl forming helical structures. This gives them the ability to form a variety of different gels at room temperature. They are widely used in the food and other industries as thickening and stabilizing agents. All carrageenans are high-molecular-weight polysaccharides made up of repeating galactose units and 3,6 anhydrogalactose (3,6-AG), both sulfated and nonsulfated. The units are joined by alternating alpha 1–3 and beta 1–4 glycosidic linkages.

There are three main commercial classes of carrageenan:

- Kappa forms strong, rigid gels in the presence of potassium ions; it reacts with dairy proteins. It is sourced mainly from *Kappaphycus alvarezii*.
- Iota forms soft gels in the presence of calcium ions. It is produced mainly from *Eucheuma denticulatum*.
- Lambda does not gel, and is used to thicken dairy products. The most common source is *Gigartina* from South America.

The primary differences that influence the properties of kappa, iota, and lambda carrageenan are the number and position of the ester sulfate groups on the repeating galactose units. Higher levels of ester sulfate lower the solubility temperature of the carrageenan and produce lower strength gels, or contribute to gel inhibition (lambda carrageenan).

Many red algal species produce different types of carrageenans during their developmental history. For instance, the genus *Gigartina* produces mainly kappa carrageenans during its gametophytic stage, and lambda carrageenans during its sporophytic stage. All are soluble in hot water, but in cold water, only the lambda form (and the sodium salts of the other two) are soluble. When used in food products, carrageenan has the EU additive E-number E407 or E407a when present as "processed eucheuma seaweed". Carrageenan is mainly composed of dietary fibre, which balances the nutrition better. Technically Carrageenan is considered a Dietary Fibre.

In parts of Scotland (where it is known as (An) Cairgean in Scottish Gaelic) and Ireland (variety used is Chondrus Crispus known in Irish Gaelic variously as carraigín [little rock], fiadháin [wild stuff], clúimhín cait [cat's puff], mathair an duilisg [mother of seaweeds], ceann donn [red head]), it is known as Carrageen Moss it is boiled in milk and strained, before sugar and other flavourings such as vanilla, cinnamon, brandy, or whisky are added. The end-product is a kind of jelly similar to pannacotta, tapioca, or blancmange.

Production

Although carrageenans were introduced on an industrial scale in the 1930s, they were first used in China around 600

B.C. (where *Gigartina* was used) and in Ireland around 400 A.D. Carrageen gelatin can be prepared at home using the traditional recipe found in Diderot's encyclopedie and used for centuries. 5oz rinsed Irish moss is cooked with 8 quarts of water for 10 minutes, stirred as it boils. Hard water should be mixed with 1/2 oz of borax. Two quarts of cold water are rapidly added to the hot brew, and after the mixture has cooled it is strained through a cloth. It is then cooled for 24 hours and becomes gelatinous.

The largest producer of industrial carrageenan currently is the Philippines, where cultivated seaweed produces about 80% of the world supply. The most commonly used sources are *E. cottonii* (*Kappaphycus alvarezii, K.striatum*) and *E. spinosum* (*Eucheuma denticulatum*), which together provide about three-quarters of the world production. These grow from the sea surface to a depth of about 2 metres. The seaweed is normally grown on nylon lines strung between bamboo floats, and it is harvested after three months or so, when each plant weighs approximately 1 kg. The *E. cottonii* variety has been reclassified as *Kappaphycus cottonii* by Maxwell Doty (1988), thereby introducing the genus *Kappaphycus*, on the basis of the phycocolloids produced (namely kappa carrageenan).

After harvest, the seaweed is dried, baled, and sent to the carrageenan manufacturer. There the seaweed is ground, sifted to remove impurities such as sand, and washed thoroughly. After treatment with hot alkali solution (e.g., 5–8% potassium hydroxide), the cellulose is removed from the carrageenan by centrifugation and filtration. The resulting carrageenan solution is then concentrated by evaporation. It is dried and ground to specification.

There are Three Types of Industrial Processing

Refined: The essential difference in the refining process is that the carrageenan is first dissolved and filtered to remove cell wall debris. The carrageenan is then precipitated from the clear solution either by isopropanol or by potassium chloride.

Mixed Processing

A hybrid technology in which seaweed is treated heterogeneously as in the semirefined process exists, but alcohol or high salt levels are used to inhibit dissolution. This process is often used on South American seaweeds and gives some of the cost benefits of semirefined processing, while allowing a wider range of seaweeds to be processed.

However, the naturally low cellulose levels in some South American seaweeds allow them to be heterogeneously processed and still be sold under the EU refined specification.

Uses

Food and Other Domestic Uses

- Desserts, ice cream, cream, milkshakes, sweetened condensed milks, and sauces: gel to increase viscosity
- *Beer:* clarifier to remove haze-causing proteins
- Pâtés and processed meats (ham, e.g.): substitute for fat, increase water retention and increase volume, or improve sliceability
- *Toothpaste:* stabilizer to prevent constituents separating
- *Fruit Gushers:* ingredient in the encapsulated gel
- *Fire fighting foam:* thickener to cause foam to become sticky
- Shampoo and cosmetic creams: thickener
- Air freshener gels
- *Marbling:* the ancient art of paper and fabric marbling uses a carrageenan mixture on which to float paints or inks; the paper or fabric is then laid on it, absorbing the colours
- *Shoe polish:* gel to increase viscosity
- *Biotechnology:* gel to immobilize cells/enzymes
- *Pharmaceuticals:* used as an inactive excipient in pills/tablets
- *Soy milk:* used to thicken, in an attempt to emulate the consistency of whole milk
- *Diet sodas:* to enhance texture and suspend flavours
- Pet food
- Personal lubricants

- Animal models of inflammation used to test analgesics (dilute lambda carrageenan solution (1–2%) injected subcutaneously causes swelling and pain)
- Veggie dogs.

Medical Uses

Studies at the Laboratory of Cellular Oncology, National Cancer Institute, Bethesda, Maryland, United States of America suggest that carrageenans might function as a topical microbicide. Laboratory studies revealed a broad spectrum antiviral activity of carrageenan .

HSV

There are indications a carrageenan-based gel may offer some protection against HSV-2 transmission by binding to the receptors on the herpes virus, thus preventing the virus from binding to cells. Research has shown a carrageenan-based gel effectively prevented HSV-2 infection at a rate of 85% in a mouse model. Some personal and condom lubricants are made with carrageenan, and several of these products (such as Divine) were found to be potent HPV inhibitors in the study (though others that listed carrageenan in their ingredients were not).

HPV

Laboratory studies have shown carrageenans are extremely potent inhibitors of HPV infection *in vitro* and in animal challenge models. Clinical trial results announced at the 2010 International Papillomavirus Conference held in Montreal, Canada indicate a carrageenan-based personal lubricant called Carraguard is effective for preventing HPV infection in women. The clinical results suggest the use of carrageenan-based personal lubricant products, such as Divine No 9 and BIOglide may likewise be effective for preventing HPV infection.

HIV

A phase 3 clinical trial by the Population Council examined whether a carrageenan-based product known as Carraguard was effective as a topical microbicide for blocking HIV infection in women. The trial ran from 2004 to 2007, with more than 4,000 South African women completing the study, but found no statistical difference in infection between those having used the lubricant and those not having used the lubricant. The trial did provide information about usage patterns, however, and showed that the gel is safe at least—not increasing infection any more than the baseline or causing significant side-effects. As such, they expect to use it as a stable delivery vehicle for experimental antiretrovirals in future studies.

Concurrent studies in macaques found the same carrageenan gels used in clinical trials to be effective against SIV challenge. This was in direct contrast with *in vitro* findings, where the compound was found to enhance HIV and SIV infections in various assays. Although compliance was believed to be one issue in clinical versus animal trials, the high viscosity and controlled nature of animal-viral inoculations (atraumatic introduction of virus using a French catheter) may be why the latter animal study observed a positive outcome.

Common Cold

Clinical studies have shown carrageenans are the first substances found to be active against common cold viruses (and not just the symptoms of the cold). Marinomed Biotechnologie, an Austrian company, conducted a study in 2010 showing that their nasal spray, which contains Carrageenan, was effective as a prophylactic treatment against the viral cause (rather than treating the symptoms) of the common cold.

Health Concerns

Starting in the late 1960s, research has linked food-grade carrageenan to gastrointestinal disease in laboratory animals, including ulcerative colitis-like disease, intestinal lesions and ulcerations and colon cancer.

In 1981, two scientists wrote in the British medical journal The Lancet: "we wish to reiterate our warning in relation to the long-term dangers in the continued use of carrageenan in our food."

Recent research, funded in part by the National Institutes of Health, has identified how carrageenan's unique chemical structure may trigger an immune response in the body; this immune response in turn may lead to inflammation of the intestines. Chronic, low-grade inflammation is a precursor to more serious disease, including colon cancer.

A 2012 study showed that mice given food-grade carrageenan in the diet exhibited impaired glucose tolerance and increased insulin resistance.

The Joint FAO/WHO expert committee on food additives states that, "based on the information available, it is inadvisable to use carrageenan or processed eucheuma seaweed in infant formulas".

There is evidence from studies performed on rats, guinea pigs, and monkeys that indicates that food-grade carrageenan may cause ulcerations in the gastro-intestinal tract and promote (though not initiate) the development of gastro-intestinal cancer.

In the 21st century similar experiments have been performed using human tissue showing similar results, but none have been definitive.

Poligeenan is produced from carrageenan when subjected to high temperatures and acidity. The average carrageenan molecule weighs more than 100,000 Da, while poligeenans have a molecular weight of less than 50,000 Da. Degraded carrageenan, or poligeenan, is routinely used by scientists to induce inflammation, in order to test anti-inflammation drugs on laboratory animals. A scientific committee working on behalf of the European Commission has recommended that the amount of degraded carrageenan be limited to a maximum of 5% (which is the limit of detection) of total carrageenan mass. Upon testing samples of foods containing high molecular weight carrageenans, results showed that many samples were contaminated with low molecular weight (degraded) carrageenan. Based on research showing that it leads to colon cancer in laboratory animals, the International Agency for Research on Cancer recognizes degraded carrageenan as a "possible human carcinogen."

Organizations including The Cornucopia Institute, a non-profit food and farm policy research organization, requested that carrageenan be withdrawn from inclusion among acceptable ingredients in organic foods by the USDA.

A study published in 2006 indicates that carrageenan induces inflammation in human intestinal epithelial cells in tissue culture through a BCL10-mediated pathway that leads to activation of NFkappaB and IL-8. Carrageenan may be immunogenic due to its unusual alpha-1,3-galactosidic link, which is part of its disaccharide unit structure. Consumption of carrageenan may have a role in intestinal inflammation and possibly, inflammatory bowel disease, since BCL10 resembles NOD2, mutations of which are associated with genetic proclivity to Crohn's Disease.

Carrageenan is reported to interfere with macrophage activity. The health concerns for Carrageenan have led to a worldwide increase in prices because of the greater regulation surrounding its use as a food additive. This has also led to large price increase for non-food uses, especially as an artistic material. One artificial alternative in many applications is methyl cellulose, which is also regulated as a food additive, but has greater market availability in America in both FDA and non-FDA approved forms.

Grades

There are two basic grades of carrageenan: refined carrageenan (RC) and semi refined carrageenan (SRC). In the United States, RC and SRC are both labeled as carrageenan. In the European Union, RC is designated by the E number E-407, and SRC is E-407a. RC has a 2% maximum for acid insoluble material and is produced through an alcohol precipitation process or potassium chloride gel press process. SRC contains much a higher level of cellulosic content and is produced in a less complex process. Indonesia, The Philippines and Chile are three main sources of raw material and extracted carrageenan.

Chondrus Crispus

Chondrus crispus — commonly called Irish moss or carrageen moss (Irish *carraigín*, "little rock") — is a species of red algae which grows abundantly along the rocky parts of the Atlantic coast of Europe and North America.

In its fresh condition this protist is soft and cartilaginous, varying in color from a greenish-yellow, through red, to a dark purple or purplish-brown. The principal constituent is a mucilaginous body, made of the polysaccharide carrageenan which comprises about 55% by weight. The organism also consists of nearly 10% protein and about 15% mineral matter, and is rich in iodine and sulfur. When softened in water it has a sea-like odour and because of the abundant cell wall polysaccharides it will form a jelly when boiled, containing from 20 to 100 times its weight of water.

Description

Chondrus crispus is a relatively small red alga, reaching up to a little over than 20 cm in length. It grows from a discoid holdfast and branches four or five times in a dichotomous, fan-like manner. The morphology is highly variable, especially the broadness of the thalli. The branches are 2–15 mm broad, firm in texture and dark reddish brown in color bleaching to yellowish in sunlight.

The gametophytes often show a blue iridescence and fertile sporophytes show a spotty pattern. *Mastocarpus stellatus* (Stackhouse) Guiry is a similar species which can be readily distinguished by its strongly channelled and often somewhat twisted thallus. The cystocarpic plants of *Mastocarpus* show reproductive papillae quite distinctively different from *Chondrus*. When washed and sun-dried for preservation, it has a yellowish, translucent, horn-like aspect and consistency.

Distribution

Chondrus crispus is common all around the shores of Ireland and Great Britain and can also be found along the coast of Europe including Iceland, the Faroe Islands western Baltic Sea to southern Spain. It is found on the Atlantic coasts of Canada and recorded from California in the United States to Japan. However, any distribution outside the Northern Atlantic needs to be verified. There are also other species of the same genus in the Pacific Ocean, for example, *C. ocellatus* Holmes, *C. nipponicus* Yendo, *C. yendoi* Yamada *et* Mikami, *C. pinnulatus* (Harvey) Okamura and *C. armatus* (Harvey) Yamada *et* Mikami.

Ecology

Chondrus crispus is found growing on rock from the middle intertidal zone into the subtidal zone.

Uses

Chondrus crispus is an industrial source of carrageenan, which is commonly used as a thickener and stabilizer in milk products such as ice cream and processed foods, including lunch meat. In Europe, it is indicated as E407 or E407b. It may also be used as a thickener in calico-printing and for fining beer or wine. Irish moss is frequently mixed with *Mastocarpus stellatus* (*Gigartina mammillosa*), *Chondracanthus acicularis* (*G. acicularis*) and other seaweeds with which it is associated in growth.

Carrageenan and agar-agar are also used in Asia for gelatin-like desserts, such as almond jelly. Presently, the major source of carrageenan is tropical seaweeds of the genera *Kappaphycus* and *Eucheuma*.

In parts of Scotland (where it is known as *(An) Cairgean* in Scottish Gaelic) and Ireland, it is boiled in milk and strained, before sugar and other flavourings such as vanilla, cinnamon, brandy or whisky are added. The end-product is a kind of jelly similar to pannacotta, tapioca, or blancmange. Similarly, in Trinidad and Tobago *Gracilaria* spp is boiled with cinnamon and milk to make a thick drink called Irish Moss that is believed to be an aphrodisiac.

In Venezuela it has been used for generations as a home remedy for sore throat and chest congestion, boiled in milk and served with honey before bed.

Irish moss is commonly used as a clarifying agent in the process of brewing (beer), particularly in homebrewing. A small amount is boiled with the wort, attracting proteins and other solids, which is then removed from the mixture after cooling.

Life History

Chondrus crispus undergoes an alternation of generation life cycle common in many species of algae. There are two distinct stages: the sexual haploid gametophyte stage and the asexual diploid sporophyte stage. In addition, there is a third stage- the carposporophyte, which is formed on the female gametophyte after fertilization. The male and female gametophytes produce gametes which fuse to form a diploid carposporophyte, which forms carpospores, which develops into the sporophyte. The sporophyte then undergoes meiosis to produce haploid tetraspores (which can be male or female) that develop into gametophytes. The three stages (male, female and sporophyte) are difficult to distinguish when they are not fertile; however, the gametophytes often show a blue iridescence.

Scientific Interest

Chondrus crispus is, compared to most other seaweeds, well-investigated scientifically. It has been used as a model species to study photosynthesis, carrageenan biosynthesis, and stress responses. The nuclear genome was sequenced in 2013. It was first example of complete genome sequence of a red macroalga. The genome size is 105 Mbp and is coding for 9,606 genes. It is characterised by relatively few genes with very few introns. The genes are clustered together, with normally short distances between genes and then large distances between groups of genes.

Chapter 4

Floating Plants

Each species listed below is linked to a written description which is in turn linked to photographs and line drawings.

Azolla

Azolla (mosquito fern, duckweed fern, fairy moss, water fern) is a genus of seven species of aquatic ferns in the family Salviniaceae. They are extremely reduced in form and specialized, looking nothing like conventional ferns but more resembling duckweed or some mosses.

Ecology

Azolla filiculoides (Red Azolla) is the only member of this genus and of the family Azollaceae in Tasmania. It is a very common native aquatic plant in Tasmania. It is particularly common on farm dams and other still water bodies. The plants are small (usually only a few cm across) and float, but can be very abundant and form large mats. The plants are typically red, and have very small water repellent leaves. *Azolla* floats on the surface of water by means of numerous, small, closely overlapping scale-like leaves, with their roots hanging in the water.

They form a symbiotic relationship with the cyanobacterium *Anabaena azollae*, which fixes atmospheric nitrogen, giving the plant access to the essential nutrient. This has led to the plant being dubbed a "super-plant", as it can readily colonise areas of freshwater, and grow at great speed - doubling its biomass every two to three days. The only known limiting factor on its growth is phosphorus, another essential mineral. An abundance of phosphorus, due for example to eutrophication or chemical runoff, often leads to *Azolla* blooms.

The nitrogen-fixing capability of *Azolla* has led to *Azolla* being widely used as a biofertiliser, especially in parts of southeast Asia. Indeed, the plant has been used to bolster agricultural productivity in China for over a thousand years. When rice paddies are flooded in the spring, they can be inoculated with *Azolla*, which then quickly multiplies to cover the water, suppressing weeds. The rotting plant material releases nitrogen to the rice plants, providing up to nine tonnes of protein per hectare per year. *Azolla* are also serious weeds in many parts of the world, entirely covering some bodies of water. The myth that no mosquito can penetrate the coating of fern to lay its eggs in the water gives the plant its common name "mosquito fern".

Most of the species can produce large amounts of deoxyanthocyanins in response to various stresses, including bright sunlight and extremes of temperature, causing the water surface to appear to be covered with an intensely red carpet. Herbivore feeding induces accumulation of deoxyanthocyanins and leads to a reduction in the proportion of polyunsaturated fatty acids in the fronds, thus lowering their palatability and nutritive value.

Azolla cannot survive winters with prolonged freezing, so is often grown as an ornamental plant at high latitudes where it cannot establish itself firmly enough to become a weed. It is not tolerant to salinity; normal plants can't survive in greater than 1-1.6‰, and even conditioned organisms die in over 5.5‰ salinity.

Reproduction

Like all ferns, sexual reproduction leads to spore formation, but *Azolla* sets itself apart from other members of its group by producing two kinds. During the summer months, numerous spherical structures called sporocarps form on the undersides of the branches. The male sporocarp is greenish or reddish and looks like the egg mass of an insect or spider. It is two millimeters in diameter, and inside are numerous male sporangia. Male spores (microspores) are extremely small and are produced inside each microsporangium. Curiously, microspores tend to adhere in clumps called massulae. Female sporocarps are much smaller, containing one sporangium and one functional spore. Since an individual female spore is considerably larger than a male spore, it is termed a megaspore.

Azolla has microscopic male and female gametophytes that develop inside the male and female spores. The female gametophyte protrudes from the megaspore and bears a small number of archegonia, each containing a single egg. The microspore forms a male gametophyte with a single antheridium which produces eight swimming sperm. The barbed glochidia on the male spore clusters are assumed to cause them to cling to the female megaspores, thus facilitating fertilization.

Human Use

Food

In addition to its traditional cultivation as a bio-fertilizer for wetland paddy (due to its ability to fix nitrogen), azolla is finding increasing use for sustainable production of livestock feed. Azolla is rich in proteins, essential amino acids, vitamins and minerals. Studies describe feeding azolla to dairy cattle, pigs, ducks, and chickens, with reported increases in milk production, weight of broiler chickens and egg production of layers, as compared to conventional feed.

One FAO study describes how azolla integrates into a tropical biomass agricultural system, reducing the need for inputs. Azolla has also been suggested as a food stuff for human consumption. However, no long term studies of the healthiness of eating Azolla have been made on humans.

Companion Plant

Azolla has been used, for at least one thousand years [1], in rice paddies as a companion plant, because of its ability to both fix nitrogen, and block out light to prevent any competition from other plants, aside from the rice, which is planted when tall enough to poke out of the water through the azolla layer.

Larvicide

As an additional benefit to its role as a paddy biofertilizer, *Azolla* spp. have been used to control mosquito larvae in rice fields. The plant grows in a thick mat on the surface of the water, making it more difficult for the larvae to reach the surface to breathe, effectively choking the larvae.

Climatic Paleontology

A study of Arctic climatology reported that azolla may have had a significant role in reversing an increase in greenhouse effect that occurred 55 million years ago that caused the region around the north pole to turn into a hot tropical environment. This research conducted by the Institute of Environmental Biology at Utrecht University claims that large dense patches of *Azolla* growing around freshwater lakes formed by the climate change eventually consumed enough carbon dioxide for the greenhouse effect to reverse.

Floating Leaf Flower

Floating-leaf plants are rooted in the lake bottom, but their leaves and flowers float on the water surface. Water lilies are a well-known example. Floating leaf plants typically grow in protected areas where there is little wave action.

- Duckweeds and watermeal

Common Names: Lesser duckweed, duck's meat, water lentil.

Location: Ponds and quiet backwaters of lakes and streams; some duckweeds are often found near creek inlets or in ditches. Rarely will duckweed become overly abundant on lakes and large ponds exposed to wind and heavy wave action. Watermeal is often found growing with duckweed.

Description: Tiny, free-floating green plants. Watermeal resembles small grains floating on the water surface; no roots are present. Duckweed typically consists of a leaf or cluster of leaves with small roots that hang down into the water; leaves and stem are not distinguishable from each other.

Hints to Identify: From a distance, duckweed is often mistaken for algae; it may form a thick, green blanket on the water surface. Duckweed is not interconnected, as is filamentous algae. Watermeal resembles green cornmeal floating in the water.

Importance of Plants: Provide food for waterfowl and marsh birds; support insects that fish eat. However, may shade out larger, submerged plants.

Management Strategy: See DNR regulations. Because duckweed and watermeal reproduce rapidly and are small, effective control is difficult. If control is absolutely necessary, try using aquatic herbicides.

Duckweeds, or water lentils, are aquatic plants which float on or just beneath the surface of still or slow-moving fresh water bodies and wetlands. Also known as "bayroot", they arose from within the arum or aroid family (Araceae), and therefore, often are classified as the subfamily Lemnoideae within the Araceae. Classifications created prior to the approximate end of the twentieth century tend to classify them as a separate family, Lemnaceae.

These plants are very simple, lacking an obvious stem or leaves. They consist of a small "thalloid" or plate-like structure only a few cells thick that float on or just under the water surface, with or without simple rootlets.

Reproduction is mostly by asexual budding, but occasionally three tiny "flowers" consisting of two stamens and a pistil are produced, by which sexual reproduction occurs. Some view this "flower" as a pseudanthium, or reduced inflorescence, with three flowers that are distinctly either female or male and which are derived from the spadix in Araceae. Anatomical research regarding the mechanics of this process has not been completed, or remains ambiguous due to considerable evolutionary reduction of these plants from their earlier relatives.

The flower of duckweed genus *Wolffia* is the smallest known flower in the world, measuring merely 0.3 mm long. The fruit produced through this occasional sexual reproduction is a *utricle*, and a seed is produced in a sac containing air that facilitates flotation.

Gymnostachys

Gymnostachys is a monotypic genus in the family Araceae. The single species is *Gymnostachys anceps* (Settler's Twine or Boorgay), native to rainforest and wet sclerophyll forest in New South Wales and Queensland in Australia.

Gymnostachys is kept to its own subfamily Gymnostachydoideae due to its unique characteristics that include an unusually structured flowering shoot and linear leaves with parallel venation.

Orontioideae

Orontioideae is a subfamily of flowering plants in the Araceae family. The subfamily consists of three genera namely, *Lysichiton* (one to two species of skunk cabbage from North America and Asia), *Orontium* (golden club, one living species from eastern North America and two extinct species known from fossils), and *Symplocarpus* (several species of skunk cabbage from North America and Asia). Characteristics of Orontioideae include medium sized pollen grains and subterranean stems. Species in the subfamily have a base chromosome number of X=13.

Lemnoideae (duckweeds)

The genera of duckweeds are: *Spirodela*, *Lemna*, *Wolffiella*, and *Wolffia*.

Duckweed genome sizes have a 10-fold range (150 to 1500 MB), potentially representing diploids to octaploids. The ancestral genus of *Spirodela* has the smallest genome size (150 MB, similar to *Arabidopsis thaliana*), while the most derived genus, *Wolffia*, contains plants with the largest genome size (1500 MB). DNA sequencing has shown that *Wolffiella* and *Wolffia* are more closely related than the others. *Spirodela* is at the basal position of the taxon, followed by *Lemna*, *Wolffiella*, and *Wolffia*, which is the most derived.

Spirodela

Spirodela is a genus of aquatic plant. It is one of several genera containing plants commonly called duckweed. *Spirodela* species are members of the Araceae under the APG II system. They were formerly members of the Lemnaceae.

Spirodela species are free-floating thalli, 2-5 plants may remain connected to each other. Plants are green, but may have a red or brown underside. Multiple roots (7 to 12) emerge from each thallus. *Spirodela* is larger (10 mm) than *Lemna* (2–5 mm, one root per thallus).

Certain species of *Spirodela* overwinter as turions, a dormant form that lacks air pockets and so sinks to the bottom of the pond. In spring turions rise to the surface and germinate to start a new population.

Spirodela often forms floating mats with related species, e.g. *Lemna* and Wolffia.

There are three species. *Spirodela punctata* is sometimes treated as *Landoltia punctata.*,

Lemna

Lemna is a genus of free-floating aquatic plants from the duckweed family. These rapidly growing plants have found uses as a model system for studies in community ecology, basic plant biology, in ecotoxicology, in production of biopharmaceuticals, and as a source of animal feeds for agriculture and aquaculture.

Taxonomy and Growth Habits

The duckweeds have been classified as a separate family, the Lemnaceae, but some researchers (the AGP II) consider the duckweeds members of the Araceae. *Lemna* species grow as simple free-floating thalli on or just beneath the water surface. Most are small, not exceeding 5 mm in length, except *Lemna trisulca* which is elongated and has a branched structure. *Lemna* thalli have a single root, which distinguishes them from related genera *Spirodela* and *Landoltia*

The plants grow mainly by vegetative reproduction: two daughter plants bud off from the adult plant. This form of growth allows very rapid colonisation of new water. Duckweeds are flowering plants, and nearly all of them are known to reproduce sexually, flowering and producing seed under appropriate conditions. Certain duckweeds (e.g. *L. gibba*) are long day plants, while others (e.g. *L. minor*) are short day plants. When *Lemna* invades a waterway, it can be removed mechanically, by the addition of herbivorous fish (e.g. grass carp) or treated with a herbicide. The rapid growth of duckweeds finds application in bioremediation of polluted waters and as test organisms for environmental studies. It is also being used as an expression system for economical production of complex biopharmaceuticals. Duckweed meal (dried duckweed) is a good cattle feed. It contains 25-45% proteins (depending on the growth conditions), 4.4% fat, and 8-10% fibre, measured by dry weight.

Assessing the Toxicity of Chemicals with Lemna

Organisation for Economic Co-operation and Development and U.S. Environmental Protection Agency (US EPA) guidelines describe toxicity testing using *Lemna gibba* or *Lemna minor* as test organisms. Both of these species have been studied extensively for use in phytotoxicity tests. Genetic variability in responses to toxicants can occur in Lemna, and there are insufficient data to recommend a specific clone for testing. The US EPA test uses aseptic technique.

The OECD test is not conducted axenically, but steps are taken at stages during the test procedure to keep contamination by other organisms to a minimum. Depending on the objectives of the test and the regulatory requirements, testing may be performed with renewal (semi-static and flow-through) or without renewal (static) of the test solution. Renewal is useful for substances that are rapidly lost from solution as a result of volatilisation, photodegradation, precipitation or biodegradation.

Production of Biopharmaceuticals

Lemna has been transformed by molecular biologists to express proteins of pharmaceutical interest. Expression constructs were engineered to cause *Lemna* to secrete the transformed proteins into the growth medium at high yield. Since the *Lemna* is grown on a simple medium, this substantially reduces the burden of protein purification in preparing such proteins for medical use, promising substantial reductions in manufacturing costs.

In addition, the host *Lemna* can be engineered to cause secretion of proteins with human patterns of glycosylation, an improvement over conventional plant gene-expression systems. Several such products are being developed, including monoclonal antibodies.

Duckweed Farming

High yields of duckweed with a high protein content can be achieved by careful control of growth conditions. Although duckweed can tolerate temperatures ranging from 6 to 33 °C, the optimal growth range is 20 to 28 °C. The acceptable pH range is 5 to 9, but better growth is obtained in the pH range of 6.5 to 7.5. A minimum water depth of 1 ft is desirable to prevent excessive temperature swings. High nitrogen levels, for example 20 mM urea, have provided a protein content in the range of 45% by weight. The water may typically contain 60 mg/L of soluble nitrogen and 1 mg/L of phosphorus. Fertiliser is required on a daily basis for optimal growth.

Duckweed can be farmed organically, with nutrients being supplied from a variety of sources, for example cattle dung, pig waste, biogas plant slurry, or other organic matter in slurry form. Because of the rapid growth of duckweed, daily harvesting is necessary to achieve optimal yields. Harvesting is done such that less than a kilogram per square metre of duckweed remains. Under optimal conditions, a duckweed farm can produce 10 to 30 tons of dried duckweed per hectare per year.

Wolffiella

Wolffiella is a genus of aquatic plants commonly called duckweeds in the Araceae family. Common names for plants in this genus include bogmat and mud-midget. They are rootless and have a keel that allows them to maintain their orientation in the water. They are small, measuring 2 to 10 millimeters in width.

Wolffia

Wolffia is a genus of 9 to 11 species which include the smallest flowering plants on Earth. Commonly called watermeal or duckweed, these aquatic plants resemble specks of cornmeal floating on the water. *Wolffia* species are free-floating thalli, green or yellow-green, and without roots. The flower is produced in a depression on the top surface of the plant body. It has one stamen and one pistil. Individuals often float together in pairs or form floating mats with related plants, such as *Lemna* and *Spirodela* species. Most species have a very wide distribution across several continents. *Wolffia* species are composed of about 40 percent protein, about the same as the soybean, making them a potential high-protein human food source. They have historically been collected from the water and eaten as a vegetable in much of Asia.

In order to identify different duckweed genomes, a DNA-based molecular identification system was developed based on seven plastid-markers proposed by the Consortium for the Barcode of Life. The *atpF-atpH* noncoding spacer was chosen as a universal DNA barcoding marker for species-level identification of duckweeds.

Research

In July 2008 the U.S. Department of Energy (DOE) Joint Genome Institute announced that the Community Sequencing Program would fund sequencing of the genome of the giant duckweed, *Spirodela polyrhiza*. This was a priority project for DOE in 2009. The research is intended to facilitate new biomass and bio-energy programs.

Duckweed is being studied by researchers around the world as a possible source of clean energy. In the United States, in addition to being the subject of study by the DOE, both Rutgers University and North Carolina State University have ongoing projects to determine if duckweed might be a source of cost-effective, clean, renewable energy. Duckweed is a good candidate as a biofuel because as a biomass it grows rapidly, has 5 to 6 times as much starch as corn, and does not contribute to global warming. Duckweed is considered a carbon neutral energy source, because unlike most fuels, it removes carbon dioxide from the atmosphere. Duckweed also functions as a bioremediator by effectively filtering contaminants such as bacteria, nitrogen, phosphates, and other nutrients from naturally occurring bodies of water, constructed wetlands and waste water.

- Floating Marsh-marigold *Rare aquatic plant Caltha palustris* (kingcup, marsh marigold) is a perennial herbaceous plant of the family Ranunculaceae, native to marshes, fens, ditches and wet woodland in temperate regions of the Northern Hemisphere.

It becomes most luxuriant in partial shade, but is rare on peat. In the United Kingdom, it is probably one of the most ancient native plants, surviving the glaciations and flourishing after the last retreat of the ice, in a landscape inundated with glacial meltwaters. Height is up to 80 centimetres (31 in) tall. The leaves are rounded to kidney-shaped, 3–20 centimetres (1.2–7.9 in) across, with a bluntly serrated margin and a thick, waxy texture. Stems are hollow.

The flowers are yellow, 2–5 cm (1–2 in) diameter, with 4-9 (mostly 5) petal-like sepals and many yellow stamens; they appear in early spring to late summer. The flowers are visited by a great variety of insects for pollen and for the nectar secreted from small depressions, one on each side of each carpel.

Carpels form into green sac-like follicles to 1 cm long, each opening to release several seeds.

Caltha palustris is a highly polymorphic species, showing continuous and independent variation in many features. Forms in the UK may be divided into two subspecies: *Caltha palustris* subsp. *palustris*, and *Caltha palustris* subsp. *minor*.

It is sometimes considered a weed in clay-like garden soils, where every piece of its root will survive and spread. In warm free-draining soils, it simply dies away.

As is the case with many members of the family Ranunculaceae, all parts of the plant are poisonous and can be irritant. Skin rashes and dermatitis have been reported from excessive handling of the plant.

Other Names and Etymology

In the UK, *Caltha palustris* is known by a variety of common names, varying by geographical region. These include marsh marigold and kingcup (the two most frequently used common names), mayflower, May blobs, mollyblobs, pollyblobs, horse blob, water blobs, water bubbles, gollins. Balfae (in Caithness) and the publican. The common name of marigold refers to its use in medieval churches at Easter as a tribute to the Virgin Mary, as in *Mary gold.*

The specific name *palustris*, Latin for "of the marsh", indicates its common habitat.

Richard Mabey, in *Flora Britannica*, describes *Caltha palustris* thus:

"Marsh-marigolds are in decline as agricultural land continues to be drained, but they are still the most three-dimensional of plants, their fleshy leaves and shiny petals impervious to wind and snow, and standing in sharp relief against the tousled brown of frostbitten grasses. Most of the plant's surviving local names - water-blobs, molly-blobs, water-bubbles - reflect this solidity, especially the splendid, rotund 'the publican' from Lancashire."

In North America *Caltha palustris* is sometimes known as cowslip. However, cowslip more often refers to *Primula veris*, the original plant to go by that name. Both are herbaceous plants with yellow flowers, but *Primula veris* is much smaller.

Caltha palustris is a plant commonly mentioned in literature, including Shakespeare:

Winking Marybuds begin

To open their golden eyes (*Cymbeline*, ii. 3).

Kingcup Cottage by Racey Helps is a children's book which features the plant. In Latvia Caltha palustris is also known as *Gundega* which is also used as a girls name which symbolizes fire. The word *Gundega* is made from 2 words - *uguns* (fire) and *dega* (burned). This refers to the burning reaction that some people experience from contact with *Caltha* sap.

Subspecies, Varieties and Cultivars

The 2006-2007 edition of the Royal Horticultural Society *Plant Finder*, a British publication which lists over 70,000 plants available in nurseries in the United Kingdom, lists the following:

- *Caltha palustris* (awarded the RHS's Award of Garden Merit)
  - *Caltha palustris* var. *alba*
  - *Caltha palustris* var. *barthei*
  - *Caltha palustris* var. *himalensis*
  - *Caltha palustris* var. *palustris*
    - *Caltha palustris* var. *palustris* 'Plena' (double flowered)
  - *Caltha palustris* subsp. *polypetala*
  - *Caltha palustris* var. *radicans*
  - *Caltha palustris* 'Flore Pleno' (a double-flowered cultivar, awarded the RHS's Award of Garden Merit)
  - *Caltha palustris* 'Semiplena' (double flowered)
  - *Caltha palustris* Trotter's form
  - *Caltha palustris* 'Yellow Giant'

  * *Caltha palustris* 'Marilyn'
  * *Caltha palustris* 'Multiplex' (double flowered)
- Small White Water-lily *Rare aquatic plant*

Nymphaea alba, also known as the European White Waterlily, White Lotus, or Nenuphar, is an aquatic flowering plant of the family Nymphaeaceae.

It grows in water from 30-150 centimeters deep and likes large ponds and lakes. The leaves can be up to thirty centimeters in diameter and they take up a spread of 150 centimeters per plant. The flowers are white and they have many small stamens inside.

The red variety which is in cultivation came from lake *Fagertärn* (Fair tarn) in the forest of Tiveden, Sweden, where they were discovered in the early 19th century. The discovery led to a large scale exploitation which nearly made it extinct in the wild before it was protected.

It is found all over Europe and in parts of North Africa and the Middle East in freshwater.

Nymphaea candida J. Presl is sometimes considered a subspecies of *N. alba* (*N. alba* L. subsp. *candida* (J. Presl) Korsh.).

It contains the active alkaloids nupharine and nymphaeine, and is a sedative and an aphrodisiac/anaphrodisiac depending on sources. Although roots and stalks are used in traditional herbal medicine along with the flower, the petals and other flower parts are the most potent.

Alcohol can be used to extract the active alkaloids, and it also boosts the sedative effects. The root of the plant was used by monks and nuns for hundreds of years as an anaphrodisiac, being crushed and mixed with wine.

In the earliest printed medical textbooks, authors would maintain this area of use, though warning against consuming large and frequent doses.

- *Spatterdock:* Yellow Pond Lily, also called Spatterdock, is a native water lily which grows in lakes, ponds, and slow-moving rivers. It is a perennial, which means it dies back in the winter, but grows again from the roots each Spring.

Yellow Pond Lily has thick leaves, which can be round or heart-shaped. They are large, growing up to 12 inches wide. The leaves are attached to a long stalk which extends down into the water and into the mud below. When the water level drops (especially in rivers, streams, or marshes affected by tides), the leaves will be above the water. When the water level rises again, the leaves will float on the water.

The flowers of this plant are yellow, with thick petals. Fruit are shaped like pods and contain many seeds.

Nuphar advena (spatterdock or yellow pond-lily) is a species of *Nuphar* native throughout the eastern United States, and just into the extreme south of Canada. It is similar to the Eurasian species *N. lutea*, and is treated as a subspecies of it by some botanists, though differing significantly in genetics.

It is Locally Naturalized in Britain

The stalks of Yellow Pond Lily are attached to a type of underground stem with roots, called a rhizome. Rhizomes are buried deep in the mud underwater. Rhizomes can multiply and grow new rhizomes. This allows Yellow Pond Lilies to grow quickly and form huge colonies.

The stalks connecting leaves and flowers to rhizomes can grow six feet long. In the Fall, leaves and flowers turn brown quickly and die. The rhizome lives on in the mud. Yellow Pond Lilies provides great cover for wildlife, especially fish, aquatic insects, snakes, turtles, frogs, crayfish, salamanders, and other water creatures. Beaver and Common Muskrats eat the rhizomes, and Beaver also eat the leaves. Waterfowl, such as Wood Ducks, Mallards, and Canada Geese eat the seeds of Yellow Pond Lily. Frogs, salamanders, and aquatic insects also use leaves and stems to attach egg masses. Leaves, or lily pads, provide resting places for terrestrial (land) insects, such as dragonflies and bees.

Medicinal and Food Uses

Spatterdock was long used in traditional medicine, with the root applied to the skin and/or both the root and seeds eaten for a variety of conditions. The seeds are edible, and can be ground into flour. The root is edible too, but can prove to be incredibly bitter in some plants.

- *Swamp smartweed: Polygonum* is a genus in the Polygonaceae family. Common names include knotweed, knotgrass, bistort, tear-thumb, mile-a-minute, and several others. In the Middle English glossary of herbs "Alphita" (ca. 1400-1425), it was known as ars-smerte. There have been various opinions about how broadly the genus should be defined. For example, Buckwheat has sometimes been included in the genus.

The genus primarily grows in northern temperate regions. They vary widely from prostrate herbaceous annual plants under 5 cm high, others erect herbaceous perennial plants growing to 3–4 m tall, and yet others perennial woody vines growing to 20–30 m high in trees. Several are aquatic, growing as floating plants in ponds. The smooth-edged leaves range from 1–30 cm long, and vary in shape between species from narrow lanceolate to oval, broad triangular, heart-shaped, or arrowhead forms. The stems are often reddish or red-speckled. The small flowers are, pink, white, or greenish, forming in summer in dense clusters from the leaf joints or stem apices.

The genus name is from the Greek *poly*, "many" and *gonu*, "knee" in reference to the swollen jointed stem. *Polygonum* species are occasionally eaten by humans, and are used as food plants by the larvae of some Lepidoptera species. Most species are considered weedy, especially in moist soils in the USA.

Distribution and Uses

Several species can be eaten cooked, for example during famines. The variety *Polygonum cognatum*, known locally as "madimak", is regularly consumed in central parts of Turkey.

In Chinese medicine, a *Polygonum* extract called Rèlínqîng Kçlì (íp—gn—˜'|) is used to treat urinary tract infections. Chinese medicine also uses a Polygonum multiflorum extract called Fo-Ti. Care should be taken not to confuse Polygonum with Polygonatum - an entirely different genus of plants.

Species

Between 65 and 300 species are recognised, depending on the circumscription of the genus; some botanists divide the genus into several smaller genera, including *Fagopyrum*, *Fallopia* and *Persicaria*.

The genus *Polygonella* has a number of morphological similarities with *Polygonum*, and some authors have included *Polygonella* in *Polygonum*.

- Watershield

Water-shield may refer to following aquatic plants:

- *Brasenia schreberi* (water-shield)

Brasenia

Brasenia is a genus belonging to the family Cabombaceae, consisting of one extant species widely distributed in warm temperate and tropical regions of the world. Brasenia is an perennial aquatic plant with floating, peltate leaves and rhizomatous stems.

It is identified by its bright green leaves, small purple flowers that bloom from June through September, and a thick mucilage that covers all of the underwater organs, including the underside of the leaves, stems, and developing buds. This mucilage may be an anti-herbivore defence trait, perhaps to deter snail grazing. It grows in shallow water of lakes, rivers and beaver ponds, particularly those with somewhat acidic water.

Characteristics

Brasenia exhibits wind pollination. The flowers have a two-day blooming period. On the first day, the functionally female, or pistillate flower, extends above the surface of the water and exposes the receptive stigmas.

The flower then recedes below the water surface and on the following day emerges as a functionally male, or staminate flower. It is elevated higher than on the previous day and the anther-bearing filaments are extended beyond the female carpels. The anthers dehisce, releasing the pollen, and the flower is then withdrawn below the water where the fruit develops.

Uses

Economically, Brasenia is cultivated as a vegetable in China and Japan (where it is known as junsai) and the mucilage it produces has been found to have anti-algal and anti-bacterial properties that may be useful as a natural weed control.

History

Brasenia occurred during the interglacial of Europe, but like some other aquatic species, it does not occur there now.

Name

Brasenia schreberi (syn. *B. nymphoides*, *B. peltata*) has the common name water-shield (also watershield or water shield).

The genus commemorates the surgeon and Moravian missionary Christoph Brasen (1738-1774), who was the first superintendent of the Moravian mission at Nain in Labrador.

- *Cabomba caroliniana* (fanwort; Carolina water-shield) *Cabomba caroliniana*, commonly known as Green Cabomba, is an aquatic perennial herbaceous plant native to North America. It is a Weed of National Significance in Australia.[1]

Common Names

Green Cabomba, Fanwort, Carolina Fanwort, Fish grass, Washington grass

Distribution

South America (Amazon region), eastern and west coast United States Eaten as a vegetable in some areas.

Ecological Aspects

Grows rooted in the mud of stagnant to slow flowing water, including streams, smaller rivers, lakes, ponds, sloughs, and ditches. In some States in the United States it is now regarded as a weed. Fanwort stems become brittle in late summer, which causes the plant to break apart, facilitating its distribution and invasion of new waterbodies. It produces by seed but vegetative reproduction seems to be its main vehicle for spreading to new waters. Growth of 50 mm (2 in) a day has been reported in Lake Macdonald in Queensland, Australia.

Large numbers of plants are sent from Florida to the rest of the U.S. for commercial use. Fanwort is also grown commercially in Asia for export to Europe and other parts of the world. Small-scale, local cultivation occurs in some area and aquarists are probably responsible for some introductions.

Description

Fanwort is a submersed, sometimes floating, but often rooted, freshwater perennial plant with short, fragile rhizomes. The erect shoots are upturned extensions of the horizontal rhizomes. The shoots are grass green to olive green or sometimes reddish brown. The leaves are of two types: submersed and floating. The submersed leaves are finely divided and arranged in pairs on the stem. The floating leaves, when present, are linear and inconspicuous, with an alternate arrangement. They are less than 1/2 inch (13 mm) long and narrow (less than 1/ 4 inch or 6 mm). The leaf blade attaches to the centre, where there is a slight constriction. The flowers are white and small (less than 1/2 inch (13 mm) in diameter), and are on stalks which arise from the tips of the stems.

Cultivation

A nutriment rich water on the soft side. Additional carbon dioxide will encourage strong growth. The easiest of the Cabomba species to grow in the aquarium but needs a strong bright light. Plant in groups. Can grow fast in good conditions. If the water circulation is too strong or with some fish (which like to nibble at it), because of its brittle stems, filters may become clogged.

White Water Lily: The white water lily is a perennial plant that often form dense colonies. The leaves arise on flexible stalks from large thick rhizomes. The leaves are more round than heart-shaped, bright green, 6 to 12 inches in diameter with the slit about 1/3 the length of the leaf. Leaves usually float on the water's surface. Flowers arise on separate stalks, have brilliant white petals (25 or more per flower) with yellow centres. The flowers may float or stick above the water and each opens in the morning and closes in the afternoon. The flowers are very fragrant. White water lily can spread from seeds or the rhizomes.

Submerged portions of all aquatic plants provide habitats for many micro and macro invertebrates. These invertebrates in turn are used as food by fish and other wildlife species (e.g. amphibians, reptiles, ducks, etc). After aquatic plants die, their decompostion by bacteria and fungi provides food (called "detritus") for many aquatic invertebrates. Deer, beaver, muskrat, nutria and other rodents will consume the leaves and rhizomes of white water lily, while the seeds are eaten by ducks.

Yellow Lotus: The white water lily is a perennial plant that often form dense colonies. The leaves arise on flexible stalks from large thick rhizomes. The leaves are more round than heart-shaped, bright green, 6 to 12 inches in diameter with the slit about 1/3 the length of the leaf. Leaves usually float on the water's surface. Flowers arise on separate stalks, have brilliant white petals (25 or more per flower) with yellow centres. The flowers may float or stick above the water and each opens in the morning and closes in the afternoon. The flowers are very fragrant. White water lily can spread from seeds or the rhizomes.

Submerged portions of all aquatic plants provide habitats for many micro and macro invertebrates. These invertebrates in turn are used as food by fish and other wildlife species (e.g. amphibians, reptiles, ducks, etc). After aquatic plants die, their decompostion by bacteria and fungi provides food (called "detritus") for many aquatic invertebrates. Deer, beaver, muskrat, nutria and other rodents will consume the leaves and rhizomes of white water lily, while the seeds are eaten by ducks.

Description

American lotus is an emergent aquatic plant. It grows in lakes and swamps, as well as areas subject to flooding. The roots are anchored in the mud, but the leaves and flowers emerge above the water's surface. The petioles of the leaves may extend as much as 2 m (6.6 ft) and end in a round leaf blade 33–43 cm (13–17 in) in diameter. Mature plants range in height from 0.8 to 1.5 m (2.6 to 4.9 ft).

Flowering begins in late spring and may continue into the summer. The specific name means "yellow" in Latin and refers to the flowers, which may be white to pale yellow. The flowers measure 18–28 cm (7.1–11 in) in diameter and have 22-25 petals.

Range

The native distribution of the species is the southeastern United States, Mexico, Honduras, and the Caribbean. It was apparently distributed northwards in the United States by Native Americans who carried the plant with them as a food source.

Uses

This plant has a large tuber that is used as a food source. This may be the plant called "macoupin" in Miami-Illinois. The seed is also edible and is known as "alligator corn". It is widely planted in ponds for its foliage and flowers. American Lotus has established itself as a weed in some areas, spreading via creeping rhizomes and seeds. This species has been crossed with *N. nucifera* to create many hybrids. Seeds may be propagated by scarifying the pointed tip of the seed with a file then soaking in water, or by division of established plants.

Other Media

Disney's character Princess Tiana (Disney) wears as her iconic princess dress, a gown fashioned from a yellow Nelumbo lutea.

Floating Ferm

Salvinia molesta D. S. Mitchell is a floating fern native to South America that in the last half of the twentieth century spread widely throughout the tropics and subtropics, moved in part by the trade in ornamental plants for fish tanks and ponds. It forms dense mats over lakes and slow moving rivers and causes large economic losses and a wide range of ecological problems to native species and communities. It is of interest in the United States because of its recent establishment in east Texas.

Nature of Damage

Economic damage. Mats of *S. molesta* (referred to hereafter as salvinia) impede access to and use of waterways for commercial and recreational purposes and degrade waterside aesthetics. Mats reduce habitats for game birds, limit access to fishing areas, and probably alter fisheries, all with negative economic consequences. Salvinia can clog water intakes and interfere with agricultural irrigation, water supply, and electrical generation. It provides habitats for vectors of human disease with serious socioeconomic impacts.

In developing countries, the impact of salvinia can be devastating because weed mats block the use of waterways for transportation, cutting off access to important services, farm lands, and hunting grounds. The harm from salvinia mats to fisheries also can be very significant to communities dependent on fish for local consumption (sometimes as the main source of protein) or in areas where fish sales are the main source of cash income (Bennett, 1966; Thomas and Room, 1986). Salvinia also is a weed of paddy rice that reduces production by competing for water, nutrients and space (Anon., 1987).

Ecological damage. The ability to grow very quickly (Cary and Weerts, 1983; Mitchell and Tur, 1975; Mitchell, 1978/9; Room, 1986) and blanket water bodies makes salvinia an aggressive and competitive weed. Initially, salvinia forms a single layer over water, but with continued growth the mats become multi-layered and can reach up to 1 m in thickness (Thomas and Room, 1986). Thick mats support other colonizing plants, and the high biomass and stability of such mats make them difficult to dislodge and destroy (Storrs and Julien, 1996).

Plants and animals dependent on open water to gain sunlight, oxygen, and space for sustenance and growth, or for landing, fishing, nest building, or mating, are displaced by dense salvinia infestations. Water under mats of salvinia has a lower oxygen concentration (due to reduced surface area of water available for oxygenation, inhibition of photosynthesis by submerged plants, and consumption of dissolved oxygen by decaying salvinia), higher carbon dioxide and hydrogen sulphide concentrations, lower pH, and higher temperatures than nearby open water (Mitchell, 1978; Thomas and Room, 1986). Through high growth rates and slow decomposition rates, salvinia reduces the concentration of nutrients that would otherwise be available to primary producers and organisms that depend on them (Sharma and Goel, 1986; Storrs and Julien, 1996). Mats of salvinia provide ideal habitat for *Mansonia* mosquitoes, a principal vector of rural elephantiasis in Sri Lanka (Pancho and Soerjani, 1978), and for other mosquito species involved in the transmission of encephalitis, dengue fever, and malaria (Creagh, 1991/92). Two species of Mansonia that occur in the United States, *Mansonia dyari* Belkin and *Mansonia titillans* (Walker), have been implicated in the transmission of St. Louis encephalitis and Venezuelan equine encephalitis, respectively (Lounibos et al., 1990).

Extent of losses. The most detailed assessment of costs caused by salvinia was conducted in Sri Lanka using 1987 as the base year (Doeleman, 1989). Paddy rice losses, fishing losses, other losses (power generation, transport, washing and bathing, etc.), health costs, abatement costs, and economic benefits were considered. No environmental costs were included, but they were recognized as important. There were no identified benefits from salvinia. Total costs associated with salvinia were estimated to be between 24.7 million and 56.7 million rupees (in Australian dollars, between 0.9 and 2.1 million) for 1987. This information was used to determine the benefits from biological control over the following 25 years. The benefits were 53 rupees or dollars per rupee or dollar invested, or 1,673 man-hours per man-hour invested.

Using this information as a guide, Room and Julien (1995) estimated that the annual benefits gained from successful biological control of salvinia worldwide were approximately $150 million U.S.

Geographical Distribution

The native range of salvinia is an area in southeastern Brazil (Forno and Harley, 1979). Its first recorded exotic establishment was in Sri Lanka in 1939 (Williams, 1956). It has since become established in India (Cook and Gut, 1971), Australia (Room and Julien, 1995), Papua New Guinea (Mitchell, 1979), Cuba, Trinidad, Guyana, Columbia (Holm et al., 1979), South Africa (Cilliers, 1991), Botswana (Edwards and Thomas, 1977), Kenya, Zambia (Mitchell and Tur, 1975), Namibia (Forno and Smith, 1999), Madagascar (Room and Julien, 1995), Ghana and Cote D'Ivoire (M. Julien, pers. obs.), Indonesia (Java, Borneo, Sulawesi), Malaysia (mainland Sabah, Sarawak) (R. Chan, pers. comm.), the Philippines (Pablico et al., 1989), Fiji (Kamath, 1979), and New Zealand (Randall, 1996).

Salvinia was first reported outside of cultivation in the United States in 1995 at a pond in southeastern South Carolina (Johnson, 1995). It was eradicated before spread occurred. It was next found in Houston, Texas, in May 1998, and then at other sites in Texas and in Louisiana during 1998. During 1999 it was found in ponds and rivers in Alabama, Arizona, California, Florida, Georgia, Hawaii, Mississippi, and Oklahoma (Jacono *et al.*, 2000). Salvinia is readily available for purchase in the United States, particularly through the Internet.

Background Information on Pest Plant

Taxonomy

The aquatic fern family Salviniaceae is placed within the order Hydropteridales and consists of a single genus, Salvinia. Ten species of *Salvinia* occur worldwide (Herzog, 1935; de la Sota, 1962, 1963, 1964, 1982; Mitchell, 1972). None are native to the United States (Jacono et al., 2000) although seven species originate in the Americas (de la Sota, 1976). *Salvinia molesta* was given recognition as a species in 1972 (Mitchell, 1972) and is grouped within the *Salvinia auriculata* complex, together with *Salvinia auriculata* Aublet, *Salvinia biloba* Raddi, and *Salvinia herzogii* de la Sota (Mitchell and Thomas, 1972). Species within this complex are characterized by the presence of divided but apically joined "basket" hairs on the abaxial surface, which produce an "egg-beater-like" appearance (de la Sota, 1962; Mitchell and Thomas, 1972; Forno, 1983). *Salvinia molesta* can be distinguished from species within the S. auriculata complex by the arrangement of sporangia, the shape of sporocarps (Mitchell and Thomas, 1972; Mitchell, 1972), and by the pattern of leaf venation (Forno, 1983). The accepted common name is salvinia, but it also is called Kariba weed, water fern, or African pyle (in Africa); giant azolla or Australian azolla (in the Philippines); and giant salvinia, water spangles, or floating fern (in the United States).

Salvinia minima Baker, the only other *Salvinia* species present in the United States also is exotic and can be distinguished by the presence of divided hairs on the abaxial leaf surface that are spreading and free at the tips.

Biology

Plant form. Salvinia is a free-floating aquatic fern with a horizontal rhizome just beneath the water surface (Bonnet, 1955; Room, 1983). Each plant is a colony of ramets. Each ramet comprises an internode, a node, a pair of floating leaves, the submerged 'root,' and associated buds. The "root" is a modified leaf that looks and functions like a root (Croxdale 1978, 1979, 1981).

Salvinia is morphologically variable, primarily in response to the level of crowding and availability of nutrients. These two factors are largely independent of one another. There are three growth forms, with a continuum among them, that are associated with the degree of crowding experienced by the plant (Mitchell and Tur, 1975).

The *primary form* occurs as isolated plants in the initial 'invading' stage of an infestation. This form has small, oval leaves less than 15 mm wide that lie flat on the water surface.

The *secondary form* occurs when plants have been growing over open water for some time, either freely or on the edge of stable mats. Internodes are longer, with larger, boat-shaped (slightly keeled) leaves that have rounded apices and are variable in size, but are normally between 20 mm and 50 mm wide. The entire lower leaf surface is in contact with the water.

The *tertiary form* occurs when plants are growing in crowded mat conditions associated with mature infestations. Internodes are short with large heart-shaped, or oblong and deeply keeled leaves up to 60 mm in width when fully opened. The undersides of adjacent leaves are in contact with each other.

Growth and reproduction. Salvinia is pentaploid, has a chromosome number of 45, and is incapable of sexual reproduction (Loyal and Grewal, 1966). Each node bears a series of up to three axillary buds that develop successively under normal growing conditions (Room, 1988), and up to six in response to damage (Julien and Bourne, 1986). The number of axillary buds that grow, the rate of growth, and plant size are largely dependent on available nutrients. Growth is apically dominant and progresses by expansion of apical and axillary buds, the latter forming branches. New plants form when older plants break apart due to senescence or damage (Room, 1983).

Factors affecting growth. Salvinia is a perennial plant with no seasonal periodicity, although changes in growth may be related to seasonal variations such as changes in temperature. Salvinia is well adapted to growth in low nutrient waters and can take up nutrients quickly when they become available

(Room and Thomas, 1986). The proportion of axillary buds that develop is correlated with the nitrogen content of the plant (Room, 1983; Julien and Bourne, 1986), and the nitrogen content increases following removal of buds by insects or other agents (Room and Thomas, 1985; Julien and Bourne, 1986; Forno and Semple, 1987). At low levels of nitrogen leaves are larger, 'roots' longer, sporocarps occur more frequently, and rhizome branching is reduced (Room, 1983; Julien and Bourne, 1986; Room, 1988; Room and Julien, 1995). The nitrogen content of salvinia ranges from 0.6 to 4.0% dry weight (Room and Thomas, 1986). The maximum rate of nitrogen uptake, calculated from rates of growth, is near 8 mg nitrogen/g dry weight of salvinia/ day or about 6,000 kg nitrogen/ha/year (Room, 1986). Actual measurements at a sewage treatment lagoon indicated an uptake of 1,580 kg nitrogen/ha/year (Finlayson et al., 1982).

The optimum temperature for growth is 30°C. Room (1986) described the effect of temperature, above and below 30°C, on relative growth rates and predicted no growth below 10°C and above 40°C. Temperature does not affect the proportion of axillary buds that expand to initiate new branches (Room, 1988). Exposure to temperatures below –3°C or above 43°C for more than two hours kills salvinia (Whiteman and Room, 1991). Leaf temperatures that exceeded 40°C and sometime approached 50°C for the hottest parts of days did not obviously affect growth, but water temperatures remained below 40°C and probably acted as a heat sink for the plants (Storrs and Julien, 1996). Plants may be killed by frost but protected parts and unfrozen buds survive. Water bodies are normally cooler than the air in summer and warmer in winter due to thermal inertia. This helps protect salvinia from temperature extremes.

Mats of salvinia can grow in water bodies with conductivities ranging from 100 μS/cm to 1,400 μS/cm (Mitchell *et al.*, 1980; Room and Gill, 1985). In water with 10% of the salinity of seawater (4,800 μS/cm), growth was reduced by 25% (Divakaran et al, 1980); at 20% salinity, growth was very slow; while at 30%, plants died after 30 minutes exposure (Room and Julien, 1995). Optimum pH for growth is 6.0 (Cary and Weerts, 1984). In the field the plant grows at pH values from 5.2 to 9.5 (Holm et al., 1977; Mitchell *et al.*, 1980). Salvinia compensates for the destruction of buds by initiating growth of dormant buds. Complete compensation occurs only when high levels of nitrogen are available (Julien and Bourne, 1986; Julien et al., 1987). Destruction of leaves (Julien and Bourne, 1988) and rhizomes (Julien and Bourne, 1986) does not induce compensatory growth. *Growth rates and density.* On Lake Kariba, Zimbabwe, numbers of leaves (ramets) doubled in eight to 17 days (Gaudet, 1973; Mitchell and Tur, 1975). In the Kakadu National Park, Australia, dry weight doubled in five to 30 days (Storrs and Julien, 1996).

Under ideal growth conditions, biomass and numbers of ramets typically double in two to three days (Mitchell and Tur, 1975; Cary and Weerts, 1983). Densities from as high as 2,500 large tertiary form ramets per m^2 (in nutrient-poor water) to 30,000 small tertiary form ramets (in nutrient-rich waters) have been noted. At these densities, natality is equaled by mortality (Room and Julien, 1995). Salvinia is 95% water by weight and biomass of living shoots can exceed 600 g/m^2 of dry weight, while biomass of living and dead shoots and 'roots' may exceed 1,600 g/m^2 of dry weight or 400 t/ha of fresh weight (Room and Julien, 1995). Fresh weight biomass in Texas varied through the year, reaching a high in October 1999 of 248 t/ha and a low of 84 t/ha in January 2000 (P. Tipping, unpub. data).

Spread. Salvinia is spread within and between aquatic systems mainly by people. It is spread accidentally when equipment or boats are moved and deliberately when it is used as a pond, aquarium, or water-garden plant or as a biological weapon (Gewertz, 1983). It is carried on animals as they move from infested water bodies (Forno and Smith, 1999). Dispersal within a water body or catchment is by wind and water currents (Room and Julien, 1995). Currents and floods wash mats away and growth is best in still or slow moving water. In its native range in southeastern Brazil, salvinia is a component of floating and emergent plant communities. Salvinia supports a variety of natural enemies (Forno and Bourne, 1984), and it normally does not form the extensive mats prevalent in its exotic range.

Analysis of Related Native Plants in the Eastern United States

The Salviniaceae are included within a monophyletic clade of heterosporous genera that also encompasses the Azollaceae and Marsileaceae (Pryer and Smith, 1998; Pryer et al., 1995; Hasebe et al., 1995), all aquatic leptosporangiate ferns. This clade includes ten species in the North American flora. The Azollaceae once were included within the Salviniaceae, but the relationship is not close and they have since been separated (Lumpkin, 1993).

Azollaceae consists of the single genus Azolla. There are three species of Azolla – *Azolla caroliniana* Willdenow, *Azolla mexicana* C. Presl, and *Azolla filiculoides* Lamarck – that are native to North America (Lumpkin, 1993). The Marsileaceae includes two genera, Marsilea and Pilularia, both of which occur in North America. Seven species within these two genera are part of the North American flora (Johnson, 1993): *Marsilea quadrifolia* Linnaeus, *Marsilea ancylopoda* A. Braun, *Marsilea oligospora* Gooding, *Marsilea mollis* B. L. Robinson and Fernald, *Marsilea macropoda* Engelmann, Marsilea vestita Hooker and Greville, and *Pilularia americana* A. Braun. *Marsilea quadrifolia*, a native of Europe and Asia, is introduced and *M. ancylopoda* is extinct, so only eight native species remain.

History of Biological Control Efforts

Area of Origin of Weed

The native range of *Salvinia molesta* includes a relatively small area (20,000 km^2) in southeastern Brazil, including the states of Sao Paulo, Paraná, Santa Catarina and Rio Grande do Sul. It occurs between the latitudes 24005' S and 32005' S; at altitudes 0 to 500 meters; and up to 200 km inland (Map 1). Salvinia occurs in natural lagoons, artificial dams, swamps, drainage canals, and along margins of rivers (Forno and Harley, 1979).

Areas Surveyed for Natural Enemies

The first surveys for potential biological control agents for *S. molesta* were conducted in Trinidad, Guyana and northeastern Brazil from 1961 to 1963 (Bennett, 1966), and in Argentina prior to 1975 (Bennett, 1975), where species in the *S. auriculata* complex other than *S. molesta* occur. At this time the true identity and the native range of S. molesta were not known. Surveys for natural enemies were conducted in Trinidad, Venezuela, Guyana, Uruguay, Paraguay, Brazil, and Argentina during 1978 to 198In 1978, the previously unknown range of *S. molesta* was identified (Forno and Harley, 1979), permitting surveys to focus on the relatively small native range of the target weed rather than the larger range of the *S. auriculata* complex (Forno and Bourne, 1984).

Natural Enemies Found

The natural enemies of *S. molesta* and the related species in the *S. auriculata* complex are listed in Forno and Bourne (1984), including species collected by Bennett (1975). Twenty-five phytophagous or possibly phytophagous species have been recorded from *S. molesta*, compared to 49 species from the four species of the S. auriculata complex. Four of these species have been used as biological control agents against S. molesta. The first three, *Cyrtobagous singularis* Hustache, *Paulinia acuminata* (De Geer), and *Samea multiplicalis* (Guenée) (identified during the early exploration [Bennett, 1966]) have not been successful control agents. The fourth, *Cyrtobagous salviniae* Calder and Sands, was found during later work (Sands, 1983) and has been extremely successful.

Host Range Tests and Results

Host range studies on the three unsuccessful species are summarized in Bennett (1966), Sankaran and Ramaseshiah (1973), Sands and Kassulke (1984, 1986), and Knopf and Habeck (1976).

Host range tests to assess feeding by *C. salviniae* were carried out in Australia on 46 species from six families of Pteridophyta (ferns), eleven families of Monocotyledons, and sixteen families of Dicotyledons (Forno et al., 1983). (This weevil was later found to be a new species and subsequently descrbed as *C. salviniae* Calder and Sands).

Test plants were exposed to mature weevils in three replicates in choice tests. Adult feeding occurred on *Pistia stratiotes* L., but the insect was unable to reproduce on that species. Minor leaf feeding was observed in choice tests on *Ipomea batatas* (L.) Lam. (sweet potato) when the leaves were held in contact with water, an abnormal condition.

Adults failed to feed on I. batatas in no-choice tests in a non-aquatic situation and died within seven days. Host specificity tests indicated that this weevil was restricted to *S. molesta*. It has never been observed attacking plants other than Salvinia species in the field in South America, including those that grew in association with S. molesta such as water fern (*Azolla* sp.), waterhyacinth (*Eichhornia crassipes* [Mart.] Solms-Laubach), and waterlettuce (*P. stratiotes*) (Forno et al., 1983). Importantly, this weevil has not been found to attack any other plants even when huge populations were starving following population crashes of salvinia.

Releases Made

The grasshopper P. *acuminata*, collected from *S. auriculata* in Trinidad, was released in Zimbabwe (in 1969 and again in 1971), Kenya and Zambia (1970), Botswana (1971 and 1975), Sri Lanka (1973 and 1978), India (1994), and Fiji (1975). It failed to establish in Botswana, Kenya, and Sri Lanka and does not provide control in the countries where it established.

The weevil *C. singularis*, collected from *S. auriculata* in Trinidad, was released in Botswana (in 1971 and 1976), Zambia (1971), and Fiji (1976). It is established in each country but is not providing control.

The moth *S. multiplicalis*, also collected from *S. auriculata* in Trinidad, was released in Zambia (1970), Botswana (1972), and Fiji (1976). It failed to establish in Fiji and Zambia and does not control the weed in Botswana (Julien and Griffiths, 1998).

It was later collected from *S. molesta* in Brazil and released in Australia during 1981, where it established widely but failed to provide control (Room et al., 1984; Forno, 1987).

Releases of *C. salviniae* from *S. molesta* in southeastern Brazil were made first in Australia in 1980. This weevil now controls the weed in most tropical and subtropical areas and in some temperate climates.

It has been released in 15 countries and controls the weed in at least 12 of these. In Cote D'Ivoire, insufficient time has elapsed to measure success. Information on project results is not available for the Philippines or Indonesia. Map 1 shows the countries that have or have had serious salvinia problems and those where biological control has been successful.

Table: *The Status of Releases of Cyrtobagous Salviniae Calder and Sands for Each Country and the Date of Initial Release (Modified from Julien and Griffiths, 1998)*

Country	***Initial Release Date***	***Status***
Australia	1980	Control in tropical and subtropical areas; some control in temperate areas
Botswana	Spread from Namibia	Control in 1 to 5 years
Cote D'lvoire	1998	Established and spreading
Fiji	1991	Successful control
Ghana	1996	Control
India	1983	Control at Bangalore and Kerala
Indonesia	1997	Status is unknown
Kenya	1990	Control except where affected by herbicide
Malaysia	1989	Control where released. Needs redistribution
Namibia	1984	Good control
Papua New Guinea	1982	Good control
Philippines	1989	Established on Panay. Impact unknown
Republic of South Africa	1985	Successful control within 2 years
Sri Lanka	1986	Successful control
Zambia	1990	Excellent control
Zimbabwe	1992	Good control within 2 years

History of Biological Control Efforts in the Eastern United States

A weevil found attacking *S. minima* in Florida was identified as *C. singularis* (Kissinger, 1966), but Calder and Sands (1985) listed Cyrtobagous specimens from Florida as *C. salviniae* and did not consider the range of *C. singularis* to include North America. When *S. molesta* was found in the United States and biological control was considered, further morphological examination of weevils from Florida suggested that they were *C. salviniae* (C. O'Brien, pers. comm.). This weevil was collected from *S. minima* in Florida by scientists from the USDA, ARS Invasive Plant Research Laboratory and released at sites in Liberty, Bridge City, and Toledo Bend Reservoir in Texas, and at Salter Creek in Louisiana during 1999.

When another USDA laboratory planned to release an Australian population of *C. salviniae*, a molecular comparison of the D2 gene was made between Florida and Australian material to provide a means of distinguishing between weevils from the two sources. Several base pair differences were found but the taxonomic significance of this is not yet clear. Further studies utilizing molecular, morphological and bioassay methods are planned. Until these differences are explained, further release of the Florida population will be suspended. No release of Australian material has been done yet.

Biology and Ecology of Key Natural Enemies

Salvinia weevils, *Cyrtobagous salviniae* and *C. singularis* (Coleoptera: Curculionidae) The genus *Cyrtobagous* Hustache was originally thought to be monotypic, containing only the species *C. singularis*. This weevil was known to feed on various South American Salvinia species and had been used unsuccessfully as a biological control agent for salvinia during the 1970s (Julien and Griffiths, 1998). Discovery of the native range of S. molesta in the late 1970s allowed surveys for natural enemies to concentrate on the target weed.

A weevil thought to be *C. singularis* was collected from S. molesta and subsequently released in Australia, where it controlled the weed. Comparative studies determined that this was a new species, subsequently named *C. salviniae*. Differences in the biology between the two species explained why one species failed to cause significant damage to the weed while the other proved to be an excellent control agent. Larvae of C. salviniae tunnel within the rhizomes causing them to disintegrate. Larvae also tunnel in the buds and adults eat buds, thus suppressing growth and vegetative propagation of this sterile weed. Larvae and adults of C. singularis feed on leaves and other tissues but don't affect the rhizomes or meristems. This research provided a classic example of the importance of careful taxonomic study of both the weed and the insects for successful biological control (Thomas and Room, 1986). Other important differences between the two species are that C. salviniae has a higher intrinsic rate of increase, lays seven times more eggs, and oviposits with greater frequency. Furthermore, this species' oviposition is less affected by changes in the nutrient status of the host plant, and larval and pupal survival rates are higher.

C. salviniae. The adult male of C. salviniae (1.8 x 0.9 mm) is slightly smaller than the female (2.2 x 1.2 mm). Newly emerged adults are brown, darkening to black in about five days. Detailed descriptions are given in Calder and Sands (1985) of the features that distinguish this species from C. singularis. Adults are found on or beneath young leaves, on or inside the developing leaves or among 'roots'. When under water, adults respire by means of an air bubble (called a plastron) that adheres to their ventral surface (Forno et al., 1983). Multiple matings occur five to 26 days after emergence. At 25.5°C, oviposition begins after six to 14 days. Eggs (0.5 x 0.24 mm) are laid singly in cavities excavated by adults in lower leaves, developing leaves, rhizomes, and 'roots.' At 25.5°C, females lay one egg every two to five days for at least 60 days (Forno et al.., 1983). Eggs hatch in 10 days at this temperature. Newly emerged larvae (1 mm) are white. They feed initially on 'roots' in or on the small buds, and later inside rhizomes, completing three instars in approximately 23 days (Forno et al., 1983). Development rate is dependent on temperature and the nutrient status of the host plant, larval development taking 13.4 days at 31°C on 'high' nitrogen plants.

Larvae prefer to tunnel in young rhizomes and more tunneling occurs if plants are low in nitrogen. Larvae do not survive below 16.3°C (Sands et al., 1983). Pupation occurs in a cocoon (2 x 2.6 mm), which is woven from 'root hairs' and attached underwater to the 'roots,' rhizomes or leaf bases. At 25.5°C, pupae require 12.6 days for full development. Pupal duration is not affected by plant quality (Forno et al., 1983; Sands et al., 1983). Oviposition does not occur below 21°C, and eggs fail to hatch below 20°C or above 36°C. Adults feed between 13°C and 33°C (Forno et al., 1983). The lower lethal temperature at which 50% of the adult population would be expected to die is -5.2°C (Reaney 1999).

Population densities of C. salviniae are capable of reaching 300 adults and 900 larvae per m^2, levels estimated as necessary for control (Room, 1988, 1990; Room and Thomas, 1985).

Weevils in the genus Cyrtobagous were first recorded from the United States in Florida at the Archbold Biological Station (Highlands County) in 1962 (Kissinger, 1966).

It is assumed that these weevils were accidentally introduced from South America, because of the lack of any earlier U.S. records and the adventive status of its host plant, S. minima. Kissinger (1966) considered the Florida weevils to be C. singularis, but this was before C. salviniae was recognized as a separate species. Calder and Sands (1985) later classified the Florida specimens as C. salviniae, but noted that the C. salviniae from S. minima in Florida were significantly smaller than those from S. molesta in Brazil. Based solely on morphological features, the weevils from Florida (later released in Texas) seem to be C. salviniae (C. O'Brien, pers. comm.). However, recent DNA assessments suggest that the Florida material differs from C. salviniae from Australia in some respects (Goolsby et al., 2000). Whether these differences imply separate species status is under investigation.

C. singularis. The biology of *C. singularis* is less well known and it is presumably similar to *C. salviniae*. For morphological differences between adults of this species and those of *C. salviniae*, and for larvae, May and Sands (1986). Feeding differences between the species are outlined in Sands and Schotz (1985), and other differences in life history and intrinsic rates of increase are discussed in Sands et al. (1986). Adults are small (2 to 3 mm), black, sub-aquatic weevils that reside on or beneath leaves.

While under water they respire using a plastron. Adults preferentially feed on apical leaves but also on the second to fifth pair of leaves, buds, and petioles (Sands and Schotz, 1985). Eggs are laid singly in cavities made by females in leaves. Unlike *C. salviniae*, whose larvae feed internally, those of *C.*

singularis feed on the outer surfaces of submerged buds, rhizomes, and petioles. Feeding results in bud destruction, but not rhizome disintegration, and plants retain their capacity for regrowth (Sands and Schotz, 1985). Field population densities of *C. singularis* do not exceed 50 adults per m^2 (Schlettwein, 1985), a level that is insufficient to significantly damage salvinia (Room, 1990). This insect has not been a useful biological control agent (Julien and Griffiths, 1998).

Waterlettuce moth, Samea multiplicalis *(Lepidoptera: Pyralidae)*

Larvae and adults of *S. multiplicalis* are very similar to the closely related species, *Niphograpta* (Sameodes) *albiguttalis* (Warren) (waterhyacinth moth, Pyralidae). Centre *et al.* (1982) give characters to separate larvae of these species. Samea multiplicalis was originally described from Brazil where it was observed feeding on waterhyacinth. It is widely distributed throughout warmer regions of North and South America. In Florida, it is most commonly found on waterlettuce but also is present on A. caroliniana and S. minima. It is occasionally abundant on small waterhyacinth plants, feeding within inflated leaf petioles (Knopf and Habeck, 1976). This species was introduced into Australia for biological control of both *S. molesta* and *P. stratiotes* (Sands and Kassulke, 1984).

Adults are 6.5 to 10.5 mm long (Sands and Kassulke, 1984), tan, with brown and cream markings on both fore and hind wings. Females lay approximately 300 eggs during their four-to-seven-day life span (Knopf and Habeck, 1976; Sands and Kassulke, 1984; Taylor, 1984). Moths prefer to lay eggs on undamaged salvinia plants with high nitrogen content (Taylor and Forno, 1987).

Most eggs are laid singly among the epidermal plant hairs on the lower surfaces of waterlettuce leaves, on the upper surfaces of Salvinia leaves, or lodged between the leaves of Azolla species. Eggs hatch in about four days at 26°C. Larvae may feed from within a refugium (made of silk and plant hair) attached to the external leaf surface or, for waterlettuce, within galleries in the leaves. Larvae periodically extend the refugium to reach fresh leaf material (Knopf and Habeck, 1976). Larger larvae feed on the buds of plants, often killing the growing apex. Larvae also will eat mature waterlettuce fruits and consequently destroy enclosed seeds.

On salvinia, temperature and plant quality interact to determine rates of insect growth, number of larval instars, fecundity, and survival (Taylor and Sands, 1986; Taylor, 1984, 1988, 1989). Optimum temperature for development for all stages is 28 to 30°C. Fecundity is greatest at 20 to 22°C and egg survival highest at 25 to 26°C (Taylor, 1988). Larvae complete development in 17 to 35 days on salvinia (Sands and Kassulke, 1984; Taylor, 1984). Development on waterlettuce and other salvinia species is described in Bennett (1966) and Knopf and Habeck (1976).

Pupation occurs within a silken cocoon. On waterlettuce, this cocoon is usually formed within the spongy portion of a leaf but on S. molesta it is constructed among old leaves. Pupal development requires four to seven days at 28°C on waterlettuce and S. minima (Knopf and Habeck, 1976) and eight to nine days at 26°C on *S. molesta* (Sands and Kassulke, 1984). Total development (egg to adult) requires 24.6 (Knopf and Habeck, 1976) to 42 days (Sands and Kassulke, 1984).

Intrinsic rates of increase are highest in autumn and decline in summer and winter (Room et al., 1984; Taylor, 1988). Levels of parasitism (24%) and disease in Australian populations on salvinia do not explain the seasonal variation in population growth rates (Semple and Forno, 1987); rather, field population densities are strongly determined by temperature and the nutritional quality of the plant (Taylor, 1988).

Salvinia grasshopper, Paulinia acuminata (Orthoptera: Pauliniidae)

This semi-aquatic grasshopper is adapted to living on floating mats of host plants where humidity is constantly high. Paulinia acuminata can complete its development on *Salvinia* spp., *P. stratiotes*, *Azolla* sp., and *Hydromystria* sp. and will feed on a range of other plant species (Bennett, 1966; Sands and Kassulke, 1986). It feeds on all species in the *S. auriculata* complex from Trinidad to northern Argentina and Uruguay (Bennett, 1966). Eggs of *P. acuminata*, in an ootheca, are attached underwater to the undersides of leaves and hatch in 19 to 21 days. Out of water, eggs fail to hatch (Bennett, 1966).

There are five or six nymphal stages, six being common when temperatures are lower (Thomas, 1980). Eggs take 17 to 20 days to hatch; six nymphal instars complete development after 47 days; the pre-ovipositional period takes eight to 10 days; and duration from egg to adult is 67 days on average (Sands and Kassulke, 1986).

Females live 50 days or more and lay 200 or more eggs. Adults usually are brachypterous but macropterous forms occasionally occur in the field and are frequent in laboratory cultures (Bennett, 1966). Adults and nymphs feed on all plant parts above the water level when food is scarce. Grasshoppers prefer to feed on new growth and oviposition is reduced when the weed is matted (Mitchell and Rose, 1979). Under extremely hot conditions adults shelter in the water with their heads exposed (Thomas, 1980). In the field at Lake Kariba, where the mean annual temperature was 24 to 28°C, it was estimated that P. acuminata could complete three generations per year (Thomas, 1980).

A significant reduction in salvinia on Lake Kariba occurred in 1972/3, following the release of *P. acuminata* in 1970. However, the decline of the weed at this location has been attributed to other factors such as nutrient stress on the weed (Mitchell and Rose, 1979; Marshall and Junor, 1981). Chisholm (1979) demonstrated that densities of more than 85 P. acuminata per m^2 feeding for 24 days were required to reduce production of new leaves. On Lake Kariba during the period when salvinia was a problem the maximum grasshopper density recorded was 27 per square meter (Marshall and Junor, 1981), suggesting that the insect might not have been a primary cause of decline in the weed. Sands and Kassulke (1986) noted that field populations in Africa of 45 to 54 P. acuminata per m^2 grazed up to 87% of leaves without affecting apical or lateral growth or killing plants.

Evaluation of Project Outcomes

In the United States, *S. molesta* is a new problem so evaluations are premature. As mentioned previously, *C. salviniae*, a naturalized species on *S. minima* in Florida, was released in Texas in 1999. It will take another year to determine if establishment has occurred.

Recommendations for Future Work

Currently, three release sites and three insect-free control sites are being monitored monthly in water bodies in eastern Texas and western Louisiana infested with S. molesta using standard protocols. In addition, molecular techniques are being utilized to identify and compare at least six salvinia species, including those outside of the S. auriculata complex. This will provide a means of identifying Salvinia species should any new invasions occur in the United States or elsewhere. Similar molecular testing of Cyrtobagous sp. collected from different salvinia species may provide insights to the taxonomy of this group and their associated host ranges.

We anticipate that the same favourable results as obtained in at least 12 countries on three continents can be repeated in the United States, namely, reductions in the density and abundance of S. molesta, to acceptable levels. Reductions may range from local extinctions to maintenance of small populations of the weed along the fringes of ponds, lakes, and rivers.

Neptune's Crown Floating Plant

A wonderful floating plant that will grow across your water garden pool. The plant will develop beautiful maroon floating bulbs similar to the floating water hyacinth that are striking against the green foliage and produces dainty white flowers. Excellent in reducing algae in the pool. Makes a most attractive display in the water garden.

Water Hyacinth

The seven species of water hyacinth constitute the genus *Eichhornia*. Water hyacinth is a free-floating perennial aquatic plant(or hydrophyte) native to tropical and sub-tropical South America. With broad, thick, glossy, ovate leaves, water hyacinth may rise above the surface of the water as much as 1 meter in height. The leaves are 10–20 cm across, and float above the water surface. They have long, spongy and bulbous stalks. The feathery, freely hanging roots are purple-black. An erect stalk supports a single spike of 8-15 conspicuously attractive flowers, mostly lavender to pink in colour with six petals. When not in bloom, water hyacinth may be mistaken for frog's-bit (*Limnobium spongia*).

One of the fastest growing plants known, water hyacinth reproduces primarily by way of runners or stolons, which eventually form daughter plants. Each plant can produce thousands of seeds each year, and these seeds can remain viable for more than 28 years. The common water hyacinth (*Eichhornia crassipes*) are vigorous growers known to double their population in two weeks.

Invasiveness as an Exotic Plant

Water hyacinth has been widely introduced in North America, Asia, Australia, Africa and New Zealand. They can be found in large water areas such as Louisiana, or in the Kerala Backwaters in India. In many areas it, particularly *E. crassipes*, is an important and pernicious invasive species. First introduced to North America in 1884, an estimated 50 kilograms per square metre of hyacinth once choked Florida's waterways, although the problem there has since been mitigated.

When not controlled, water hyacinth will cover lakes and ponds entirely; this dramatically impacts water flow, blocks sunlight from reaching native aquatic plants, and starves the water of oxygen, often killing fish (or turtles).

The plants also create a prime habitat for mosquitos, the classic vectors of disease, and a species of snail known to host a parasitic flatworm which causes schistosomiasis (snail fever). Directly blamed for starving subsistence farmers in Papua New Guinea, water hyacinth remains a major problem where effective control programs are not in place. Water hyacinth is often problematic in man-made ponds if uncontrolled, but can also provide a food source for gold fish, keep water clean and help to provide oxygen to man-made ponds.

Water hyacinth often invades bodies of water that have been impacted by human activities. For example, the plants can unbalance natural lifecycles in artificial reservoirs or in eutrophied lakes that receive large amounts of nutrients. *Eichhornia crassipes*, the Common water hyacinth, has become an invasive plant species on Lake Victoria in Africa after it was introduced into the area in the 1980s.

Control: Water hyacinth can be controlled using three methods:

Chemical Control

The application of herbicides for controlling water hyacinth has been carried out for many years and it has been found that there is a good success rate when dealing with small infestations. A main concern when using herbicides is the environmental and health related effects, especially where people collect water for drinking and washing.

Physical Control

Physical control is performed by land based machines such as bucket cranes, draglines, or boorm or by water based machinery such as aquatic weed harvester, dredges, or vegetation shredders. Mechanical removal is seen as the best short-term solution to the proliferation of the plant. A project on Lake Victoria in Africa used various pieces of equipment to chop, collect, and dispose of 1500 hectares of water hyacinth in a 12 month period. It is however costly and requires the use of both land and water vehicles, but it took many years for the lake to become in poor condition and reclamation will be a continual process.

Biological Control

As chemical and mechanical removal is often too expensive and ineffective, researchers have turned to biological control agents to deal with water hyacinth. The effort began in the 1970s when USDA researchers released three species of weevil known to feed on water hyacinth into the United States, *Neochetina bruchi*, *N. eichhorniae*, and the water hyacinth borer *Sameodes albiguttalis*. Although meeting with limited success, the weevils have since been released in more than 20 other countries. However, the most effective control method remains the control of excessive nutrients and prevention of the spread of this species.

May 2010 the USDA's Agricultural Research Service released *Megamelus scutellaris* as a biological control insect for the invasive waterhyacinth species.

Megamelus scutellaris is a small planthopper insect native to Argentina. Researchers have been studying the effects of the biological control agent in extensive host-range studies since 2006 and concluded that the insect is highly host-specific and will not pose a threat to any other plant population other than the targeted water hyacinth. Researchers also hope that the biological control will be more resilient than existing biological controls to the herbicides that are already in place to combat the invasive water hyacinth.

Uses

Since the water hyacinths are so prolific, harvesting them for industrial use serves also as a means of environmental control. In the Philippines, Thailand and Vietnam the water hyacinth's stems are used as a braiding material and a source of fibres.

Strings of dried fibres are woven or interlinked together to form a braid or cord used for making bags, footwear, wreaths, hats, vases, Christmas lanterns, and more decorative materials. Dried stems are used for baskets and furniture. Water hyacinth fibres are used as a raw material for paper.

Since the plant has abundant nitrogen content, it can be used a substrate for biogas production and the sludge obtained from the biogas.

However, due to easy accumulation of toxins, the plant is prone to get contaminated when used as feed.

The plant is extremely tolerant of, and has a high capacity for, the uptake of heavy metals, including Cd, Cr, Co, Ni, Pb and Hg, which could make it suitable for the biocleaning of industrial wastewater.

In addition to heavy metals, *Eichhornia crassipes* can also remove other toxins, such as cyanide, which is environmentally beneficial in areas that have endured gold mining operations.

Water hyacinth removes arsenic from arsenic contaminated drinking water. It may be a useful tool in removing arsenic from tube well water in Bangladesh.

Water hyacinth is also observed to enhance nitrification in waste water treatment cells of living technology. Their root zones are superb micro-sites for bacterial communities.

Water hyacinth is a common fodder plant in the third world especially Africa though excessive use can be toxic. It is high in protein (nitrogen) and trace minerals and the goat feces are a good source of fertilizer as well.

Salvinia

Salvinia, a genus in the family Salviniaceae, is a floating fern named in honour of Antonio Maria Salvini, a 17th Century Italian scientist. The genus was published by Séguier, in Pl. Veron. 3: 52. 1754. About ten species exist. *Salvinia* is related to the other water ferns, including the Mosquito fern *Azolla*. Recent sources include both *Azolla* and *Salvinia* in Salviniaceae, although each genus was formerly given its own family.

Salvinia, like the other ferns in order Salviniales are heterosporous, producing spores of differing sizes. However, leaf development in *Salvinia* is unique. The upper side of the floating leaf, which appears to face the stem axis, is morphologically abaxial (J. G. Croxdale 1978, 1979, 1981).

The plants pose a particular hindrance on certain lakes, having choked off much of the water in Lake Bistineau near Doyline in Webster Parish, Louisiana. Other problems have persisted in a second Webster Parish site, Caney Lakes Recreation Area.

Distribution

Mostly tropical, North America, Mexico, West Indies, Central America, South America, Eurasia and Africa, including Madagascar.

Description

Small, floating aquatics with creeping stems, branched, bearing hairs but no true roots. Leaves in whorls of 3, with 2 leaves green, sessile or short-petioled, flat, entire, and floating, 1 leaf finely dissected, petiolate, rootlike, and pendent. Submerged leaves bearing sori that are surrounded by basifixed membranous indusia (sporocarps); sporocarps of 2 types, bearing either megasporangia that are few in number (ca. 10), each with single megaspore, or many microsporangia, each with 64 microspores. Spores of 2 kinds and sizes, both globose, trilete. Megagametophytes and microgametophytes protruding through sporangium wall; megagametophytes floating on water surface with archegonia directed downward; microgametophytes remaining fixed to sporangium wall. The small, hairlike growths, known as microgametical follicles, are not known to have any productive function, and are currently a biological mystery.

Economic Uses

Giant salvinia (*Salvinia molesta*) is a commonly introduced invasive weed in warm climates. It grows rapidly and forms dense mats over still waters. It is native to South America. A tiny weevil, *Cyrtobagous salviniae*, has been used successfully to control giant salvinia.

S. *natans* is commonly used in aquariums as a decorative floating plant.

Water Lettuce

Water lettuce is a floating plant. Experts disagree as to whether water lettuce is native to the U.S.: it has been present in Florida since as early as 1765 when the explorer, William Bartram, described and drew the plant in Lake George. This floating plant commonly forms large infestations which prevent boating, fishing and other uses of lakes and rivers. Water lettuce occurs in lakes, rivers and canals, occasionally forming large dense mats.

As its name implies, water lettuce resembles a floating open head of lettuce. Water lettuce has very thick leaves. The leaves are light dull green, are hairy, and are ridged. There are no leaf stalks. Water lettuce roots are light-colored and feathery. Its flowers are inconspicuous.

Pistia

Pistia is a genus of aquatic plant in the arum family, Araceae. The single species it comprises, *Pistia stratiotes*, is often called water cabbage, water lettuce, Nile cabbage, or shellflower. Its native distribution is uncertain, but probably pantropical; it was first discovered from the Nile near Lake Victoria in Africa. It is now present, either naturally or through human introduction, in nearly all tropical and subtropical fresh waterways. The genus name is derived from the Greek word ðéóôüò (*pistos*), meaning "water," and refers to the aquatic nature of the plants.

Description

It is a perennial monocotyledon with thick, soft leaves that form a rosette. It floats on the surface of the water, its roots hanging submersed beneath floating leaves. The leaves can be up to 14 cm long and have no stem. They are light green, with parallel veins, wavy margins and are covered in short hairs which form basket-like structures which trap air bubbles, increasing the plant's buoyancy. The flowers are dioecious, and are hidden in the middle of the plant amongst the leaves. Small green berries form after successful fertilization. The plant can also undergo asexual reproduction. Mother and daughter plants are connected by a short stolon, forming dense mats.

Ecology

The growth habit can make it a weed in waterways. It is a common aquatic weed in the United States, particularly in Florida where it may clog waterways. It has the potential to reduce the biodiversity of a waterway. Mats of *Pistia* block gas exchange at the air-water interface, reducing the oxygen in the water and killing fish. They also block light, killing native submerged plants, and alter immersed plant communities by crushing them.

Mosquitoes of the genus *Mansonioides* complete their life cycle only in the presence of aquatic plants such as *Pistia*, laying their eggs under the leaves. The emerging larvae fall into the water within 24 hours and stay attached to the *Pistia* root (which is rich with air sacs) with the help of a serrated siphon tube for respiration and develop into pupa. The pupa is also attached to the pistia root with the serrated piercing siphon tube. The egg to adult mosquito development is completed within 7 days.

Control

Pistia can be controlled by mechanical harvesters that remove the water lettuce from the water and transport it to disposal on shore. Aquatic herbicides may also be used. Two insects are also being used as a biological control. Adults and larvae of the South American weevil *Neohydronomous affinis* feed on *Pistia* leaves, and the larvae of moth *Spodoptera pectinicornis* from Thailand. Both are proving to be useful tools in the management of *Pistia*.

Uses

Water lettuce is often used in tropical aquariums to provide cover for fry and small fish. It is also helpful as it outcompetes algae for nutrients in the water, thereby preventing massive algal blooms.

Variegated Water Lettuce

Variegated Water Lettuce (*P. stratiotes 'variegated'*) is a most unusual plant that resembles the water lettuce but has streaks of creamy yellow in the leaves and the leaves are more erratic. Variegation stronger in subdued light. A most unusual sight floating in the water garden pool.

Water Lettuce is a floating plant with thick, fleshy, light green ribbed leaves that form a rosette that looks like a small head of lettuce. Individual rosettes can range from 2" to 18" across, depending on variety. The leaves are covered in tiny hairs that repel water. Like the water hyacinth, water lettuce is a vigorous grower and can form large mats that choke out other plants. It should never be released in a natural body of water and is outlawed in some southern states.

Water lettuce is grown for its foliage, but it does bloom. The insignificant flowers are usually hidden down between the leaves. Being a floating plant, water lettuce obtains all it?s nutrients directly from the water. This makes it a great plant to use to combat algae. The long feathery roots also make great shelter for young fish and other wildlife.

Propagation is by division of daughter plants. These plantlets form from runners and can be broken away from mother plant.

Generally care-free, but there are a few reasons it may not thrive. Fish love to nibble at the roots, and if you have large fish, they may kill the plant or even eat the entire thing. Placing the lettuce in floating baskets or partitioning them off from the fish with rocks may help. Lettuce loves to be crowded and even though they are floating plants, they don?t like to drift around. Water lettuce also does not like to have water on its leaves. Lack of nutrients will cause water lettuce to grow poorly and turn yellow.

Lettuce is such a vigorous grower it sometimes uses up one or more nutrients in the pond. This can especially be a problem in very small ponds and ponds with very few fish. There are two ways of fertilizing your lettuce. If you only have a few, you can float them in a bucket of Miracle Grow for a few hours at a time. If you have too many lettuce to remove from the pond, you can treat the entire pond with a solution of Muriate of Potash.

Varieties: There are a couple different varieties of water lettuce that water gardeners may be interested in. The two most common are the Ruffled Water Lettuce and Jurassic Water Lettuce. ?Ruffles? has more wavy leaves and doesn?t grow as large as the other varieties. It is perfect for small container ponds. Jurassic Lettuce is a large form that can grow to be the size of a dinner plate. It doesn't seem to grow as fast as the common variety. Splash is a variegated form that stays smaller. It is still very hard to find.

Winter Care: Trying to over winter water lettuce in cold regions is usually more hassle than it's worth. They grow fast enough during the summer that it is usually easier and more cost effective to buy new ones each spring. If you want to try keep lettuce over winter it needs a warm spot that is brightly lit. A south window is best, but artificial lighting may still be necessary. Place lights 6" above the plants. Some people have reported success by putting a couple inches of compost in the bottom of the container the lettuce is in. Others actually pot the lettuce and submerge the pot so the rim is an inch or two below the water surface.

Identification

The plant resembles lettuce, hence the common name. The leaves are pale to bright green, 30 to 150 mm long and up to 80 mm wide. These grow from a short stem to form a densely crowded rosette. The fan-shaped leaves are ribbed towards the extremities, with conspicuous ovoid swellings of spongy cells on the upper surface.

Both surfaces are covered with short depressed water-repellent hairs, which give the foliage a velvety appearance. Basal leaves in the rosette may lie flat on the water surface or cluster vertically. A dense cluster of branched roots, up to 450mm long, lies under water. The flower head, shaped like a miniature Arum lily contains small inconspicuous flowers. A cup-shaped membrane separates the male and female flowers within the flower head. The fruit are 6 to 10 mm long, oval, and have few seeds. Seeds are oblong, wrinkled, disc-like and 2 mm long.

Habitats

Water lettuce requires fertile water where temperatures approach 20°C for most of the year. It is also very frost tender.

Impact to Biota and Ecosystems

Water lettuce is one of the three most serious aquatic weeds in the world. A dense surface cover of water lettuce can restrict stock access to water, increase water loss by transpiration, and impede drainage and navigational. It can also block water supply inlets, dam waterways, cause flooding, and prevent recreational use of waterways.

Dispersal Routes, Vectors, Infestation Sources

Vegetative reproduction can be extremely rapid, when stolons (runners) from a parent stem extend horizontally beneath the water surface carrying a terminal bud which grows into a new plant. This stolon production is repeated by each daughter stem.

Management

Management	
Physical Control	Control work is only to be carried out by persons authorised by the Minister of Agriculture and Fisheries. Eradication is achieved by way of hand clearance.
Disposal	
Chemical Control	
Biological Control	
Recommended Approach	

Ruffled Water Lettuce

Common Name: Water Lettuce

Latin Name: Pistia stratiotes

Hardiness: zone 10

Light: Does best with shade during the hottest part of the day, but can adapt to full sun.

Size: Individual rosettes can range from 2 to 18 across.

Water Lettuce is a floating plant from Tropical America. Its thick, fleshy, light green leaves are heavily ribbed and form a rosette that looks like a small head of lettuce. The leaves are covered in tiny hairs that repel water. Like the water hyacinth, water lettuce is a vigorous grower and can form large mats that choke out other plants. It should never be released in a natural body of water and is outlawed in some southern states.

Water lettuce is grown for its foliage, but it does bloom. The insignificant flowers are usually hidden down between the leaves. Being a floating plant, water lettuce obtains all it's nutrients directly from the water. This makes it a great plant to use to combat algae. The long feathery roots also make great shelter for young fish and other wildlife.

Water lettuce is generally care free, but there are a few reasons it may not thrive. Fish love to nibble at the roots and if they are large or there are a lot of them, they may kill the plant or even eat the entire thing. Placing the lettuce in floating baskets or partitioning them off from the fish with rocks may help.

Lettuce loves to be crowded and even though they are floating plants, they don't like to drift around. Corralling them with a hoop of plastic tubing, a hula-hoop, fishing line, or even rocks will keep them from moving around. Water lettuce also does not like to have water on its leaves.

Lack of nutrients will cause water lettuce to grow poorly and turn yellow. Lettuce is such a vigorous grower it sometimes uses up one or more nutrients in the pond. This can especially be a problem in very small ponds and ponds with very few fish. There are two ways of fertilizing your lettuce. If you only have a few, you can float them in a bucket of Miracle Grow for a few hours at a time. If you have too many lettuce to remove from the pond, you can treat the entire pond with a solution of Muriate of Potash. Flora Boost is a commercially produced potassium supplement. Lettuce and other floating plants also usually respond well to any of the micro nutrient solutions specially made for pond plants.

Another problem can occur if you try to ship water lettuce during hot weather. You will end up with a head of lettuce and no roots. Float the plants in a shade spot and they usually grow new roots very quickly. Adding a little fish emulsion to the water (not in the pond!) may help, too.

Varieties: There are a couple different varieties of water lettuce that water gardeners may be interested in. The two most common are the Ruffled Water Lettuce and Jurassic Water Lettuce. Ruffles has more wavy leaves and doesn't grow as large as the other varieties. It is perfect for small container ponds. Jurassic Lettuce is a large form that can grow to be the size of a dinner plate. It doesn't seem to grow as fast as the common variety. Splash is a variegated form that stays smaller. It is still very hard to find.

Winter Care: Trying to overwinter water lettuce in cold regions is usually more hassle than it's worth. They are cheep enough and grow fast enough during the summer that it is usually easier and more cost effective to buy new ones each spring. If you want to try keep lettuce over winter it needs a warm spot that is brightly lit. A South window is best, but artificial lighting may still be necessary. Fluorescent shop lights 6 above the plants is about right. Some people have reported success by putting a couple inches of compost in the bottom of the container the lettuce is in. Others actually pot the lettuce and submerge the pot so the rim is an inch or two below the water surface.

Propagation: division of daughter plants

X Large Salvinia

X-Large Salvinia. (*Salvinia oblongifolia*). The *X-Large Salvinia* is a very interesting plant which floats on the surface of the water and is larger in size than the *Salvinia*. It has dainty "roots" that dangle into the water which are excellent for spawning of fish. Each segment grows about 3 quarters of an inch long and remains in clusters adding a wonderful colorful addition to any pool. A very excellent grower that is rather rare and will add a special look to the water garden pool. Per portion. Prohibited in TX. Free floating aquatic fern developing from a somewhat zigzagged horizontal stem (rhizome) that floats just under the water surface. The rhizome bears opposite pairs of rectangular floating leaves and a third submersed, filamentous leaf along each node. The roots are absent, although the submersed leaf has a root-like look. Floating leaves are three times longer than wide, measuring about 2.7 to 4.2 cm long and 1.0 to 1.8 cm wide. They are positioned above a prominent keel of spongy aerenchyma tissue which aids in keeping the large leaves afloat. Floating leaves lie flat or can be folded on the mid-rib depending on the density of the population. Individual plants commonly grow 20 - 25 cm long, often resembling the size and shape of a loaf of bread. The long rectangular leaves and spongy keel are characteristic for this species. Nearly all parts of the plant, except the upper leaf surface are sparsely clothed in chestnut coloured hairs.

Common Name: Salvinia Oblongifolia - Giant Salvinia

Proper Name: Salvinia Oblongifolia

Category: Floating Fern (Large Leaf 2.5cm to 4.5cm)

Country Origin: Central South America

Temperature: 15 - 35 C

Lighting: Medium

Growth Rate: Fast

Difficulty: Easy

Position in Aquascape: Floating

Propagation: by runners

Duckweed

Duckweeds, or water lentils, are aquatic plants which float on or just beneath the surface of still or slow-moving fresh water bodies and wetlands. Also known as "bayroot", they arose from within the arum or aroid family (Araceae), and therefore, often are classified as the subfamily Lemnoideae within the Araceae. Classifications created prior to the approximate end of the twentieth century tend to classify them as a separate family, Lemnaceae.

These plants are very simple, lacking an obvious stem or leaves. They consist of a small "thalloid" or plate-like structure only a few cells thick that float on or just under the water surface, with or without simple rootlets.

Reproduction is mostly by asexual budding, but occasionally three tiny "flowers" consisting of two stamens and a pistil are produced, by which sexual reproduction occurs. Some view this "flower" as a pseudanthium, or reduced inflorescence, with three flowers that are distinctly either female or male and which are derived from the spadix in Araceae. Anatomical research regarding the mechanics of this process has not been completed, or remains ambiguous due to considerable evolutionary reduction of these plants from their earlier relatives.

The flower of duckweed genus *Wolffia* is the smallest known flower in the world, measuring merely 0.3 mm long. The fruit produced through this occasional sexual reproduction is a *utricle*, and a seed is produced in a sac containing air that facilitates flotation.

Duckweed in Various Environments

One of the more important factors influencing the distribution of wetland plants, and aquatic plants in particular, is nutrient availability. Duckweeds tend to be associated with fertile, even eutrophic conditions. Duckweed can be spread by waterfowl and small mammals, transported inadvertently on their feet and bodies, as well as by moving water. In water bodies with constant currents or overflow, the plants are carried down the water channels and do not proliferate greatly. In some locations a cyclical pattern driven by weather patterns exists in which the plants proliferate greatly during low water flow periods, then are carried away as rainy periods ensue.

Duckweed is an important high-protein food source for waterfowl and also is eaten by humans in some parts of Southeast Asia. As it contains more protein than soybeans, it is sometimes cited as a significant potential food source. The tiny plants provide cover for fry of many aquatic species. The plants are used as shelter by pond water species such as bullfrogs and bluegills. They also provide shade and, although frequently confused with them, can reduce certain light-generated growths of photoautotrophic algae. The plants can provide nitrate removal, if cropped, and the duckweeds are important in the process of bioremediation because they grow rapidly, absorbing excess mineral nutrients, particularly nitrogen and phosphates. For these reasons they are touted as water purifiers of untapped value. The Swiss *Department of Water and Sanitation in Developing Countries*, SANDEC, associated with the Swiss

Federal Institute for Environmental Science and Technology, asserts that as well as the food and agricultural values, duckweed also may be used for waste water treatment to capture toxins and for odor control, and, that if a mat of duckweed is maintained during harvesting for removal of the toxins captured thereby, it prevents the development of algae and controls the breeding of mosquitoes. The same publication provides an extensive list of references for many duckweed-related topics.

These plants also may play a role in conservation of water because a cover of duckweed will reduce evaporation of water when compared to the rate of a similar size water body with a clear surface.

Taxonomy

The duckweed have long been a taxonomic mystery, and usually have been considered to be their own family, Lemnaceae. They primarily reproduce asexually. Flowers, if present at all, are small. Roots are either very much reduced, or absent entirely. They were suspected of being related to the Araceae as long ago as 1876, but until the advent of molecular phylogeny it was difficult to test this hypothesis.

Starting in 1995 studies began to confirm their placement in the Araceae and since then, most systematists consider them to be part of that family.

Their position within their family has been slightly less clear, but several twenty-first century studies place them in the position shown below. They are not closely related to *Pistia*, however, which also is an aquatic plant in the family Araceae.

The genera of duckweeds are: *Spirodela*, *Lemna*, *Wolffiella*, and *Wolffia*.

Duckweed genome sizes have a 10-fold range (150 to 1500 MB), potentially representing diploids to octaploids. The ancestral genus of *Spirodela* has the smallest genome size (150 MB, similar to *Arabidopsis thaliana*), while the most derived genus, *Wolffia*, contains plants with the largest genome size (1500 MB). DNA sequencing has shown that *Wolffiella* and *Wolffia* are more closely related than the others. *Spirodela* is at the basal position of the taxon, followed by *Lemna*, *Wolffiella*, and *Wolffia*, which is the most derived.

In order to identify different duckweed genomes, a DNA-based molecular identification system was developed based on seven plastid-markers proposed by the Consortium for the Barcode of Life. The *atpF-atpH* noncoding spacer was chosen as a universal DNA barcoding marker for species-level identification of duckweeds.

Research

In July 2008 the U.S. Department of Energy (DOE) Joint Genome Institute announced that the Community Sequencing Program would fund sequencing of the genome of the giant duckweed, *Spirodela polyrhiza*. This was a priority project for DOE in 2009. The research is intended to facilitate new biomass and bio-energy programs.

Duckweed is being studied by researchers around the world as a possible source of clean energy. In the United States, in addition to being the subject of study by the DOE, both Rutgers University and North Carolina State University have ongoing projects to determine if duckweed might be a source of cost-effective, clean, renewable energy. Duckweed is a good candidate as a biofuel because as a biomass it grows rapidly, has 5 to 6 times as much starch as corn, and does not contribute to global warming. Duckweed is considered a carbon neutral energy source, because unlike most fuels, it removes carbon dioxide from the atmosphere.

Duckweed also functions as a bioremediator by effectively filtering contaminants such as bacteria, nitrogen, phosphates, and other nutrients from naturally occurring bodies of water, constructed wetlands and waste water.

Water Sensitive Plant

As the common name suggests, the leaves, like the land based "sensitive plant", close when they are touched. The floating (mature) stems of this tropical plant are covered in white, spongy tissue, full of air cells.

Chapter 5

Submerged Plants

Submerged plants have stems and leaves that grow entirely underwater, although some may also have floating leaves. Flowers and seeds on short stems that extend above the water may also be present. Submerged plants grow from near shore to the deepest part of the littoral zone and display a wide range of plant shapes. Depending on the species, they may form a low-growing "meadow" near the lake bottom, grow with lots of open space between plant stems, or form dense stands or surface mats.

Broad-leaf Pondweeds *(Potamogeton spp.)*

Common Names: Large-leaf pondweed, claspingleaf (Richardson) pondweed, floating-leaf pondweed, whitestem pondweed, broad-leaf cabbage, musky cabbage, bass weed.

Location: Found in lakes and streams, growing in depths up to 20 feet; often grow near drop-offs.

Description: Leaves 2 to 8 inches long; large-leaf and claspingleaf pondweeds grow below the water surface, except for the flowering stalk; floating-leaf pondweed grows below the water surface, except for their flowering stalk and large, floating leaves; all species have thin and delicate submerged leaves, tough stems that are firmly rooted, and a stiff appearance when out of the water.

Hints to Identify: Leaves alternate along the stem. Often grow in patches or beds; have small seed heads crowded into spikes, often sticking up above water from June through August. Floating-leaf pondweed——Floating leaves are slightly heart-shaped. Large-leaf pondweed——Floating leaves are oval-shaped; submerged leaves are large and wavy; plants are seldom branched. Claspingleaf pondweed——Leaves are wide and wavy with a broad base that clasps the stem; plant often branches toward tip; often confused with curlyleaf pondweed, which has small "teeth" along leaf edges, but claspingleaf pondweed has no "teeth" on leaf edges.

Importance of Plants: Broad-leaf pondweeds provide excellent habitat for panfish, largemouth bass, muskellunge, and northern pike; bluegills nest near these plants and eat insects and other small animals found on the leaves; walleyes use these pondweeds for cover.

Management Strategy: These plants are important fish habitat, so it is best to let them be. Removing them may allow less-desirable aquatic plants to move in. These plants are also highly resistant to chemical control. For small problem areas, try raking or cutting the plants.

Bushy Pondweeds and Naiads *(Najas spp.)*

Common Names: Bushy naiad,(pronounced NAY-ads) water naiad, brittle naiad, slender naiad, spiny leaf naiad.

Location: Clear water at depths of up to 20 feet.

Description: Grow entirely below water surface; have long, waving stems in deep water and are dense and bushy in shallow water. These are annual plants which must start from seed each year.

Hints to Identify: Leaves are tapered to a fine point with tiny "spines"; seeds are shiny and smooth. Bushy pondweed is sometimes confused with chara, but chara has a musky odor when crushed and bushy pondweed does not.

Importance of Plants: Entire plants are eaten by waterfowl, especially mallards; provide cover for young largemouth bass, northern pike, small bluegills and perch.

Management Strategy: Bushy pondweed has become more common in recent years; it can grow abundantly in some areas and cause problems. In most areas, it will not be a nuisance and is best left alone—removal may allow less-desirable plants to move in. When control is necessary, aquatic herbicides can be effective.

Canada Waterweed (Elodea canadensis)

Common Names: Elodea (pronounced el-oh-DEE-a), American elodea, common elodea, anacharis, Canada waterweed.

Location: Found in lakes in depths up to 10 feet, often in hard water, and near stream inlets.

Description: Grows entirely underwater, except for a small white flower that blooms during the summer; has branched stems; plants and leaves are usually a dark, grass-green color; leaves are oval-shaped and arranged in clusters of three or four around the stem.

Hints to Identify: Look for leaf clusters compacted near the tip and spaced farther apart down the stem; stems are brittle, branched, and form a large mass near the lake bottom; white flowers have three to four petals and bloom during the summer; flowers are attached to the stem by a threadlike stalk.

Importance of Plant: Provides habitat for many small aquatic animals, which fish and wildlife eat; excellent oxygen producer; attractive leaves make it a popular aquarium plant. However, dense growth of this plant can create a nuisance, and its closed, compact structure is not ideal fish habitat.

Management Strategy: Canada waterweed is often associated with nuisance conditions. If the plant is controlled by cutting or pulling, pick up all fragments—they can grow into new plants. Aquatic herbicides can be an effective control method.

Coontail (Ceratophyllum demersum)

Common Name: Hornwort.

Location: Clear-to-murky water up to 20 feet deep.

Description: Grows underwater with no roots; upper leaves may reach the surface; central hollow stem has stiff, dark-green leaves; plants may be long and sparse, but are often bushy near the tip, giving the plant a "coontail" or "Christmas tree" appearance.

Hints to Identify: Often confused with watermilfoil, but coontail leaves are spiny and forked rather than feather-like.

Importance of Plant: Many waterfowl species eat the shoots; it provides cover for young bluegills, perch, largemouth bass, and northern pike; supports insects that fish and ducklings eat. However, when growing densely, commonly causes nuisance conditions along shorelines.

Management Strategy: Coontail is important to young fish, so remove as little as possible. Use cutting or raking to reduce the amount of plants. Remove all plant fragments from the water because they can regenerate into new plants. Herbicide control can be effective.

Curly-leaf Pondweed (Potamogeton Crispus)

Common Names: Curly cabbage, crisp pondweed.

Location: Grows from the shore to depths of up to 15 feet.

Description: Leaves are somewhat stiff and crinkled, approximately 1/2-inch wide and 2 to 3 inches long; leaves are arranged alternately around the stem, and become more dense toward the end of branches; produces winter buds can be confused with claspingleaf pondweed.

Hints to Identify: Has small "teeth" visible along edge of leaf; begins growing in early spring before most other pondweeds; dies back during midsummer; the flower stalks, when present, stick up above the water surface in June; appears reddish-brown in the water, but is actually green when pulled out of the water and examined closely. Easily confused with claspingleaf pondweed, which has leaves with no "teeth" around their edges.

Importance of Plant: Provides some cover for fish; several waterfowl species feed on the seeds; diving ducks often eat the winter buds.

Management Strategy: Like Eurasian watermilfoil, curlyleaf pondweed is not native to the United States and often causes problems due to excessive growth. When control is necessary, herbicides and harvesting can be effective. Grants are available for control efforts on a lake-wide basis.

Basis for Listing

Potamogeton diversifolius is a small, delicate aquatic plant that occurs in shallow water in small lakes and large ponds.

There are currently only two records of this species from Minnesota. One is a herbarium specimen collected in Ramsey County in 1945 and the other is a herbarium specimen collected in Anoka County in 1992. Since 1990, over 1,500 lakes in central and northern Minnesota have been surveyed for aquatic plants but this species was not found. Because surveys have been less thorough in the southern part of the state and Minnesota is on the northern limit of this species range, there is still hope it will be found again. Threats to its aquatic, shallow water habitat continue to grow and include shoreline development, increased recreational activities, and general decline in water quality related to runoff from surrounding uplands. For these reasons, *P. diversifolius* was listed as an endangered species in Minnesota in 1996.

Description

Potamogeton diversifolius closely resembles two other species of pondweeds that also occur in Minnesota, the rare *P. bicupulatus* (snailseed pondweed) and the more common *P. spirillus* (coiled pondweed). It is imperative that good specimens be used for reliable identification and that a specialist verifies the results. In this species, as in *P. bicupulatus*, there are 3 rows of sculpted ridges around the rim of the tiny, disk-shaped seed: 1 ridge on the rim and 2 smaller ridges, 1 on each side, of the rim. *Potamogeton diversifolius* also has extremely fine, narrow leaves but these are slightly wider than *P. bicupulatus*, typically about 0.5 mm (0.02 in.) wide. The measurement must be taken from leaves along typical segments of the underwater stem. Floating leaves, if present, are small and oval, at most about 2 cm (0.8 in) wide and 4 cm (1.6 in.) long, often smaller. Floating leaves are variable in size and are not useful for distinguishing this species (Reznicek and Bobette 1976).

Habitat

There are not enough known locations of *P. diversifolius* to gain a good understanding of its preferred lake habitats. Preliminary indications are that it prefers clear lakes on the Anoka Sand Plain with relatively low levels of dissolved minerals (soft water). These are probably small lakes or large ponds in a gently sloping basin, or perhaps a small, quiet bay in a larger lake.

Biology / Life History

Potamogeton diversifolius in Minnesota is not well known at this time. It is a perennial herbaceous species that can overwinter and resprout from a buried rhizome. Apparently it does not produce turions, which are vegetative propagules sometimes called winter buds (Haynes and Hellquist 2000). It does produce seeds regularly, which are dispersed in water currents and possibly in the fur or feathers of aquatic animals. Owing to the delicate nature of this plant, it is generally found in calmer portions of larger lakes or in smaller lakes not subject to turbulent wave action. The best time to search for *P. diversifolius* is when the plants are in fruit from July through August.

Conservation / Management

Maintaining high water quality in selected lakes in Minnesota that support this species or other rare aquatic life is a primary challenge. Activities occurring throughout a watershed can affect the water quality of lakes. Runoff can be from adjacent sources, such as shoreline development, or from sources more distant in the watershed, such as industrial manufacturing. The cumulative impacts on lake water quality and on rare plants such as *P. diversifolius* are not fully known. Certain management activities within a lake itself also pose a possible threat. Aquatic herbicides are widely used, legally or illegally, to kill aquatic vegetation. The target of the herbicides is often invasive, non-native, aquatic plants such as *Myriophyllum spicatum* (Eurasian watermilfoil) or *P. crispus* (curly-leaf pondweed). Herbicides are also used to treat luxuriant growth of native aquatic plant species perceived by lakeshore owners as unsightly or as an obstacle to water access. It is not unusual for these chemicals to be applied with inadequate understanding or consideration of the potential negative impact to rare aquatic species.

Eurasian Watermilfoil (*Myriophyllum Spicatum*)

Eurasian watermilfoil was accidentally introduced to North America from Europe. Spread westward into inland lakes primarily by boats and also by waterbirds, it reached Midwestern states between the 1950s and 1980s.

In nutrient-rich lakes it can form thick underwater stands of tangled stems and vast mats of vegetation at the water's surface. In shallow areas the plant can interfere with water recreation such as boating, fishing, and swimming. The plant's floating canopy can also crowd out important native water plants.

A key factor in the plant's success is its ability to reproduce through stem fragmentation and runners. A single segment of stem and leaves can take root and form a new colony. Fragments clinging to boats and trailers can spread the plant from lake to lake. The mechanical clearing of aquatic plants for beaches, docks, and landings creates thousands of new stem fragments. Removing native vegetation creates perfect habitat for invading Eurasian watermilfoil.

Eurasian watermilfoil has difficulty becoming established in lakes with well established populations of native plants. In some lakes the plant appears to coexist with native flora and has little impact on fish and other aquatic animals.

Likely Means of Spread: Milfoil may become entangled in boat propellers, or may attach to keels and rudders of sailboat. Stems can become lodged among any watercraft apparatus or sports equipment that moves through the water, especially boat trailers.

Narrow-leaf Pondweeds (Potamogeton Zosteriformes, Potamogeton Foliosus, Potamogeton Pectinatus)

Common Names: Flat-stemmed, leafy, and sago pondweeds.

Location: Grow below water surface; are firmly rooted; have branched, slender stems; have grasslike, narrow leaves; may have flowers or seeds extending above water surface; are limp when out of the water.

Hints to Identify: Most narrow-leaf pondweeds have no floating leaves——it is only the seeds of some types that reach or extend above the surface. Sago pondweed has stiff, threadlike leaves that appear bushy and are alternately arranged on one stem; spreading leaves in the water resemble a fan; small, pebble-sized fruits are spaced on the stem and emerge from the water.

Importance of Plants: Provide some cover for bluegills, perch, northern pike, and muskellunge, though these fish prefer broad-leaf pondweeds; good cover for walleye; provide food for waterfowl; support aquatic insects and many other small animals that fish and ducklings eat. Swans, geese, and diving ducks such as canvasbacks favour the tubers and seeds of sago pondweed.

Management Strategy: Removing narrow-leaf pondweeds may allow less-desirable plants such as curlyleaf pondweed or Eurasian watermilfoil to move in. If control is absolutely necessary, try hand-pulling, raking, or underwater cutting. Aquatic herbicides can be an effective control method.

Northern Watermilfoil (Myriophyllum Exalbescens)

Location: Grows entirely underwater in depths from 1 to 20 feet.

Description: Dark-green, feathery leaves are grouped in fours around a hollow stem that is usually buff- or pinkish-colored; leaves are comprised of 5 to 10 pairs of leaflets.

Hints to Identify: Northern watermilfoil is often mistaken for coontail or Eurasian watermilfoil, but it does not branch at the surface as much as Eurasian watermilfoil does; northern typically has half as many leaflet pairs as Eurasian has; northern leaves are rigid when removed from the water, but Eurasian leaves are limp when out of water.

The northern species also forms winter buds (groups of small, dark, brittle leaves) in late fall and winter, but the Eurasian variety does not.

Importance of Plant: Provides cover for fish and invertebrates; supports insects and other small animals eaten by fish; waterfowl occasionally eat the fruit and foliage.

Management Strategy: When control is absolutely necessary, hand-pulling, raking, or cutting can be effective in small infestations. Collect and dispose of all fragments——they can regenerate into new plants. Treatment with aquatic herbicides can be effective.

Synonyms

Potamogeton diversifolius var. *trichophyllus*

Basis for Listing

The status of this aquatic plant in Minnesota was unclear for many years because it was known only from a single lake in Anoka County where it was discovered in 1959. Attempts to locate *Potamogeton bicupulatus* in east-central Minnesota in the late 1980s and early 1990s resulted in discoveries in 4 additional lakes. These same 4 lakes also harbor other rare or uncommon species, possibly indicating that some unique combination of environmental factors is controlling this species' distribution. All of these lakes occur in areas where residential development or other activities are disturbing native vegetation in the surrounding uplands. These findings led to the listing of *P. bicupulatus* as a state endangered species in 1996.

Additional rare plant surveys were conducted in over 1,500 lakes throughout central and north-central Minnesota in the 1990s and early 2000s. *Potamogeton bicupulatus* was found in only 9 more lakes, where it was again associated with other rare or unusual aquatic plants. It tends to occur in lakes that have clear water and relatively low levels of dissolved minerals (soft water lakes). None of the lakes in which it occurs have their surrounding uplands protected, although one site is part of a private conservation area. Residential development of lakeshores continues to be one of the greatest threats to maintaining good water quality over the long-term.

Description

A small, delicate aquatic plant, *P. bicupulatus* closely resembles another rare species of pondweed that also occurs in Minnesota, *P. diversifolius* (diverse-leaved pondweed). All Minnesota sites of *P. bicupulatus* have had fruiting plants present. Fruits, though very small, appear to the naked eye to have a noticeably bumpy surface.

This is due to their 3 rows of sculpted ridges around the rim of a tiny, disk-shaped seed: 1 ridge on the rim and 2 smaller ridges, 1 on each side of the rim. These ridges are also found in *P. diversifolius* , but distinct from most other narrow-leaved pondweeds that have a single ridge or no ridge. *Potamogeton bicupulatus* also has extremely fine, hair-thin leaves, less than 0.3 mm (0.01 in.) wide, along typical segments of the underwater stem. If present, the floating leaves are small and oval, at most about 1 cm (0.4 in.) wide and 2 cm (0.8 in.) long. Floating leaves are variable in size and not definitive in distinguishing this species from other species (Reznicek and Bobette 1976).

Habitat

The characteristics of lake sediments, water quality, and surrounding vegetation in places where *P. bicupulatus* has been found have not been fully analyzed. Preliminary indications are similar to what is observed in the field — that this plant tends to occur in clear, soft water lakes. Typical aquatic plant associates include *Brasenia schreberi* (watershield), *Eriocaulon aquaticum* (pipewort), *Myriophyllum farwellii* (Farwell's watermilfoil), *Myriophyllum tenellum* (slender watermilfoil), *Najas gracillima* (slender naiad), and *Nymphaea odorata* (American white waterlily). Plants have been found rooted in pure sand and in silty sediments along the shores of shallow lakes, and rooted to the bog edge or stranded on low boggy hummocks by receding water.

Biology / Life History

Potamogeton bicupulatus appears to be a perennial herbaceous species that survives winter as a buried rhizome. It apparently does not produce turions, which are vegetative reproductive structures sometimes called winter buds (Haynes and Hellquist 2000).

It is known to produce abundant seeds, which are dispersed on water currents and possibly in the fur or feathers of aquatic animals. Owing to the delicate nature of this plant, it is generally found in calmer portions of larger lakes or in smaller lakes not subject to turbulent wave action. The best time to search for *P. bicupulatus* is when the plant is in fruit during July and August.

Conservation / Management

It is important to maintain high water quality in the select lakes in Minnesota supporting *P. bicupulatus* or assemblages of other rare aquatic plants. Activities occurring throughout a watershed can affect the water quality of lakes. Runoff can be from adjacent sources, such as shoreline development, or from sources more distant in the watershed, such as industrial manufacturing. The cumulative impacts on lake water quality and on rare plants such as *P. bicupulatus* are not fully known. Certain management activities within a lake itself also pose a possible threat. Aquatic herbicides are widely used, legally or illegally, to kill aquatic vegetation. The target of the herbicides is often invasive, non-native, aquatic plants such as *Myriophyllum spicatum* (Eurasian watermilfoil) or *P. crispus* (curly-leaf pondweed).

Herbicides are also used to treat luxuriant growth of native aquatic plant species perceived by lakeshore owners as unsightly or as an obstacle to water access. It is not unusual for these chemicals to be applied with inadequate understanding or consideration of the potential negative impact to rare aquatic species.

Potamogeton bicupulatus has been found consistently in clear-watered lakes with relatively low levels of dissolved minerals (soft water lakes). Therefore activities or actions that alter these characteristics would likely threaten existing plants or prohibit the establishment of new occurrences of the species.

Conservation Efforts in Minnesota

Targeted searches by the Minnesota DNR County Biological Survey for rare aquatic plants in Minnesota are underway. Results of the survey and ongoing water quality monitoring will contribute to our understanding of the relationship between the condition of Minnesota's lakes and rare aquatic vegetation.

Wild Celery *(Vallisneria Americana)*

Common Names: Water celery, eelgrass, tapegrass.

Location: Lakes in depths up to 15 feet and streams; prefers semi-hard bottom such as sand covered with a thin layer of muck.

Description: Leaves are ribbon-like, dark-green, and grow below the water surface; rooted in mud; in late summer, produces a small, whitish-yellow flower, supported by a coiled stalk; often grows in beds amid pondweeds and other submerged plants.

Hints to Identify: Unbranched leaves extending from the lake bottom to the water surface; flowers (and occasionally some leaves) float on the surface; leaves are attached to a horizontal central stem right above lake bottom.

Importance of Plant: Provides shade and shelter for bluegills, young perch, and largemouth bass; choice food of waterfowl, particularly diving ducks; attracts muskrats, marsh birds, and shore birds.

Management Strategy: Because wild celery is an excellent wildlife food, it is usually best left alone. Abundant growth during July and August in shallow water may interfere with recreation. Herbicides don't work well to control this plant. Hand-pulling or raking sometimes works, though floating, uprooted plants often re-establish themselves in shallow water.

Chapter 6

General Key in Aquatic Plants

Although this guide is primarily devoted to managing or reducing the amount of aquatic vegetation present in a body of water, the reader should be aware that aquatic plants have many important functions in aquatic systems. Just as on land, plants in water are natural and essential components of the environment, and without them, life as we know it could not exist.

All plants, whether on land or in the water, utilize sunlight, carbon dioxide, and water to photosynthesize. The process of photo synthesis results in the production of new plant tissue (biomass) and oxygen.

In water, new plant biomass takes the form of either microscopic plants or larger plants called macrophytes. Why are these plants so important to the aquatic environment?

1. Microscopic plants (algae) form the base of the aquatic food chain. Another term for these plants is ìphytoplanktonî or plant plankton. They, in turn, are fed upon by zooplankton, or the microscopic animal plankton. Zooplankton are fed upon by small fish, small fish by larger fish, and so on up the food chain to humans and other top predators.
2. The macrophytes (larger algae and flowering plants) provide habitat and shelter for fish, fish food organisms, water fowl, and other wildlife. 3. Macrophytes provide food for insects, water fowl, and mammals such as muskrats and beaver. However, bass, bluegill, and catfish do not, as a rule, eat macrophytic vegetation.

4. All plants produce oxygen as they photo synthesize during the daylight hours. Photosynthesis is the major source of oxygen for aquatic animal life.
5. Rooted plants stabilize shorelines and bottom sediments; uprooting these plants can lead to shoreline erosion and increased turbidity in a body of water.
6. A diversity of aquatic plant life, both around and in the water, can add visual interest and beauty to a water body. One of the largest growing sectors of the home nursery industry is water gardening. Many people now recognize and appreciate the aesthetic and natural qualities of aquatic vegetation, whether in a backyard setting, around a retention pond in an urban area, or along the shoreline of a large lake. Because of these many benefits, some aquatic plant growth is always desirable. The goal of man agreement should never be to totally eliminate aquatic vegetation from a site.

Problems Caused by Excessive Aquatic Vegetation

Unfortunately, plant growth can get out of hand. Excessive plant growth is usually due to the fact that many of our bodies of water are shallow, which allows sunlight to penetrate to the bottom and support photosynthesis. Also, our bodies of water tend to be rich in nutrients such as nitrogen and phosphorus (substances that are needed by plants for growth). In some cases, exotic species of plants that are extremely aggressive have been introduced and have taken over large areas of aquatic habitat. Some of the problems caused by excessive aquatic plant growth are as follows:

1. Recreational activities such as swimming, fishing, and boating can be impaired and even prevented.
2. Excessive growths can lead to fish stunting and overpopulation. This occurs because the production of too much habitat prevents effective predation of small fish by larger fish.
3. Aquatic plant and algae growths can play a role in causing fish kills. This usually occurs because oxygen is taken out of the water. During the day, plants produce oxygen through photo synthesis; at night (as well as

day), they consume oxygen through respiration. If plant growth is excessive, plants at night can use up most of the oxygen in the water. In fact, fish that are stressed for oxygen often die just before dawn when the oxygen content is lowest. Sometimes they can be ob served coming to the surface and gasping for air during the early morning hours. This is a sure sign that the fish are oxygen stressed and may die. Oxygen depletion also occurs when algae and plants die and decompose. When plants die, photo synthetic production of oxygen ceases, and the bacteria and fungi, which break down (decompose) the decaying plant material, use up the oxygen in their

own respiration. Unfortunately, it is difficult to predict when a plant die-off will occur under natural conditions. Changes in water temperature, pro longed periods of cloud cover, shifts in wind pat terns, and depletion of nutrients are just some of the factors that may trigger a plant or algae die-off. Another cause of plant death and oxygen depletion is the treatment of too much vegetation with a herbicide.

Similarly, it is almost impossible to predict when or if fish are going to die because of vegetation related oxygen depletion. Fish kills typically occur in the summer or in the winter. Summer fish kills can be caused, as already described, by die-offs and de composition of algae and plant growth. Another cause can be due to a dense cover of algae or free-floating plants such as duckweed or water meal on the surface of a body of water. These surface growths reduce light penetration to the deeper waters which in turn inhibits photosynthesis and oxygen production. Anything that stirs or brings these deoxygenated waters to the surface (such as oxygen levels throughout the water column and cause a h kill. Even if fish are not immediately killed, pro longed exposure to low oxygen concentrations can lead to greater fish susceptibility to diseases and toxicants. Fish kills in winter occur when snow accumulates on ice cover. Light is blocked, thus preventing photosynthesis and oxygen production by any living plants or algae. Decomposition of plants that died in the fall causes further oxygen depletion. The fact that a site has undergone a winter fish kill may not even be recognized because the dead fish can decompose and disappear under the ice or as the ice breaks. It may not be until spring or early summer that people notice that fishing is not what it used to be. Other causes of fish kills include natural stresses on fish as they come out of the winter and into the spawning season, insecticide runoff, ammonia leakage from storage tanks, and diseases.

4. Aquatic weed growth provides quiet water areas ideal for mosquito breeding.
5. Certain algae can impart foul tastes and odors to the water.
6. Weeds impede water flow in drainage ditches, irrigation canals, and culverts and cause water to back up.
7. Deposition of weeds, as well as sediment and debris, can cause the gradual filling in of bodies of water.
8. Excessive weed growth can lower property values and decrease aesthetic appeal of a body of water.
9. Exotic aquatic plant species (e.g., Eurasian water milfoil, purple loosestrife) can invade and completely take over stands of native vegetation.

This process upsets the natural balance in an aquatic system and can have adverse effects on the animals that depend on the native vegetation for habitat and food. What are the aquatic plants that most frequently cause these problems? The following guide is intended to help you identify these plants. What is an aquatic plant? The definition used in this document is: 'Aquatic plants are those whose photosynthetically active parts are submersed permanently or at least during the growing season'. This is a restrictive definition and means that most of the plants are obligate aquatics and not seasonal or marginal emergents. Most emergent and wetland plants only flower and fruit after water levels have subsided and many have only their roots submersed in water. The list of such species is quite extensive. The list of obligate aquatic plants, found only in deep permanent water of lakes and rivers is much shorter, about 100 species. Generally the species listed here reflects the authors experience in collecting plants in lakes, those species which were regularly found are included and a few commonly encountered marginal or non-obligate species are included. There is minimal overlap with the plants treated by MacKensie as Wetland Plants. With few exceptions Juncaceae, Cyperaceae and Gramineae are treated as wetland plants, not aquatics.

There will never be universal agreement as to where to draw the line between aquatic plants and wetland plants; these are arbitrary man-made boundaries in a natural continuum. We have attempted to have minimal overlap but make sure no species were omitted in assigning species to the Aquatic or Wetland category. Obligate aquatic plants usually grow in lakes and ponds or in permanent rivers, streams and sloughs. Some may survive for a while when water levels drop and their habitats are exposed but most grow in water deep enough that they are not normally exposed. Sub-tidal marine species are also included.

Rheophytes are excluded. They grow in habitats which are periodically inundated for short periods, often by flood waters, but they are established under non-flood conditions. They are firmly attached to rocks or trees to prevent being washed away by the swift currents. Rheophytes are more prevalent in tropical torrents. In high latitudes and altitudes many normally terrestrial plants may be found partially submersed in cold water much of the year.

Bryophytes are not treated except for the floating *Riccia fluitans* and *Ricciocarpus natans*. These two species may be conspicuous in some habitats and are distinct from the remaining submersed or marginal aquatic bryophytes. The treatment of Charophytes is relatively cursory. The major emphasis is on the flowering vascular plants or 'Aquatic Macrophytes'.

Not all of the plants listed are native or established outside of cultivation in British Columbia. Some species are aquarium plants that are introduced from time to time but do not persist, or are in cultivation but do not survive for long when they escape; others are major invasive weeds in other parts of the world which have not yet been found in BC but which may be expected eventually. People introduce many aquatic plants, both inadvertently and deliberately, some of which do eventually become established.

This manual includes one species of rush, *Juncus supiniformis*, three members of the Cyperaceae, *Scirpus subterminalis*, *Scirpus lacustris* and *Dulichium arundinaceum* but no grasses. Most members of the Cyperaceae, Juncaceae and Poaceae are marginal or wetland species which are difficult for non-specialists to identify at the species level. Some species may be found in the water, especially during high water, which are not included in these keys. The grass genus *Glyceria* is often found in shallow, marginal waters.

Field observations indicate that much of the variation in obligate aquatic plants which is often given specific recognition in herbaria may not be valid in nature. Many 'herbarium species' do not reflect reality. Aquatic plants are genetically plastic and respond to changing conditions with a tremendous amount of morphological variability. These morphological variants are responses to variable environmental conditions rather than indications of genetic distinctions at the species level. The species concept preferred by the author is a pragmatic 'lumping' approach, which may better reflect these field observations. However, in the interests of uniformity, the species concepts used in the four volume series 'The Vascular Plants of British Columbia' (ISSN 0843-6452) has been used in these keys. Occasional notes may be found to indicate the authors preference for another treatment.

The following pages include a list of all the species covered by these keys, a key to groups of aquatics based on their ecology, keys to each of the ecological groups, a general key to the aquatic plants of BC including a key to the aquatic plants with finely dissected submersed leaves, keys to the families and genera identified in the general key, a set of brief notes on the distribution, abundance and habitats of each species of aquatic plant, and references which may be useful in studying some aspects of the aquatic plants of BC. Illustrations of the types of dissected underwater leaves which may be found are included. Synonyms are used very sparingly and only for species which have other long-established and well known names. They are given, after the distribution and habitat data, in [brackets]. More synonyms may be found in 'The Vascular Plants of British Columbia' which also contains extensive references to the BC flora. A virtually complete list of common names gleaned from the world-wide literature is listed. There are many common names for widespread weedy species which grow in many different countries. Where more than one common name is given the preferred name for British Columbia is given first and the remainder follow in alphabetical order. Complex common names are often found in the literature and may be separate words, compound words or hyphenated words, they may also have capital or lowercase initial letters for each word. This can lead to a great many unique combinations. These combinations have been reduced to one multiple-word, non-hyphenated name with the first letter of the first word capitalized and all the rest of the words in lowercase.

The distribution limits of species are presented using vegetation zones as defined in 'The Vascular Plants of British Columbia'. The montane zone includes all continuous forests in BC except for the coastal lowlands and some islands included in the lowland zone. The subalpine zone is defined as that area above the montane zone and below the upper limit of conifers growing as an upright tree. This is represented by a meadow and tree-clump complex in the south and shrub Salix and scattered trees in the north. The alpine is above the subalpine where trees occur only as krummholz and vegetation is of tundra form. Steppe vegetation occurs in the interior and includes sagebrush or grasslands. The various keys are not mutually exclusive and may be used together to help restrict the choices or confirm the identification of a species. Using the Ecological key may limit the number of species one has to choose from in the genus or species keys, or in the dissected leaf key. Similarly if the plant obviously has dissected underwater leaves then using the dissected leaf key is more efficient than using the general key. The Ecological key may result in only a few choices which can then be readily compared with the illustrations for identification.

Some of the keys are in hierarchical sets. The Ecological key restricts you to a group of plants and the subsequent species key identifies the individual species. The key to families restricts you to genera and species if you know the family, and the key to genera identifies species if you know the genus. If you already know the family or genus you can go directly to the key that is appropriate to the level of your knowledge of the plant in question, and not have to start each time with the general key to all the aquatic plants. Using several different keys which should include the plant in question is a good check on your ability to use the keys and on the usefulness of the keys. You should get the same identification each time; if you do not perhaps you have made some assumptions about the plant which are not true or have taken the wrong fork in one of the keys. It is also possible that one of the keys can not handle some unanticipated variation in the specimen. All identifications made by keys should be checked with descriptions and illustrations, or better yet with herbarium specimens, to verify the identification. You may have a plant which is not included in the keys. The less experience you have with the species and the keys, the more important this verification becomes.

The general key to the aquatic plants of BC initially separates aquatic plants into 7 parts, for convenience and ease of use; this initial separation is based on ecology. Part 1 identifies herbaceous, fresh water plants which float freely on, or just under, the surface of the water. Part 2 identifies fully submersed, herbaceous, fresh water plants which are rooted or attached and have at least some finely dissected leaves. Part 3 identifies fully submersed, herbaceous, fresh water plants which are rooted or attached and do not have finely dissected leaves. Part 4 identifies herbaceous, fresh water plants which are rooted and emerge above the surface of the water but have cauline leaves that are opposite, whorled or clustered (more than one at each node). Part 5 identifies herbaceous fresh water plants which are rooted and emerge above the surface of the water but have cauline leaves that are alternate (only one at each node). Part 6 identifies herbaceous fresh water plants which are rooted and emerge above the surface of the water but have leaves in basal clusters as opposed to cauline leaves. Part 7 identifies fully submersed marine or brackish plants. With experience, or if you know which type of plant you have, you can go directly to the correct key. Marginal emergent shrubs such as *Potentilla palustris* are ommitted and dealt with as wetland species.

Ecological keys group plants without regard to the species and genus. They describe where and under what conditions the plants grow, in relation to the water level, and with regard to their morphology, which is often shaped by their habitat. Some are submersed, some float on the surface and some are emergent.

A number of ecological classification systems have been proposed; all are useful for specific purposes and all have their limitations.

Some of these patterns are quite distinctive and can be readily given a name: *Myriophyllum* species are all pinnate and *Ceratophyllum* species dichotomous, but most of the other types are more complex and not readily given a simple name.

The Aquatic Plants of British Columbia

- *Alisma gramineum* J. G. Gmel.
- *Alisma plantago-aquatica* L.
- *Azolla caroliniana* Willd.
- *Azolla filiculoides* Lam.
- *Azolla mexicana* Presl.
- *Brasenia schreberi* Gmel.
- {*Cabomba caroliniana*} Gray
- *Calla palustris* L.
- *Callitriche anceps* Fern.
- *Callitriche hermaphroditica* L.
- *Callitriche heterophylla* Pursh
- *Callitriche stagnalis* Scop.
- *Callitriche verna* L.
- *Caltha natans* Pallas
- *Caltha palustris* L.
- *Ceratophyllum demersum* L.
- *Ceratophyllum echinatum* Gray
- *Chara braunii* Gm.
- *Chara canescens* Desv. and Lois.
- *Chara globularis* Thuill.
- *Chara vulgaris* L.
- *Crassula aquatica* (L.) Schoenl.
- *Dulichium arundinaceum* (L.) Britt.
- *Egeria densa* Planch.
- {*Eichhornia crassipes*} (Mart.) Solms

- *Elatine rubella* Rydb.
- *Elodea canadensis* Rich.
- {*Elodea longivaginata*} St. John
- *Elodea nuttallii* (Planch.) St. John
- *Equisetum fluviatile* L. em Ehrh.
- *Equisetum palustre* L.
- *Gratiola ebracteata* Bentham
- *Gratiola neglecta* Torr.
- *Heteranthera dubia* (Jacq.) Macmill.
- *Hippuris vulgaris* L.
- {*Hydrilla verticillata*} (L.) Royle
- *Hydrocotyle ranunculoides* L. f.
- *Hydrocotyle verticillata* Thunb.
- *Isoetes bolanderi* Engelm.
- *Isoetes echinospora* Dur.
- *Isoetes howellii* Engelm.
- *Isoetes maritima* Underw.
- *Isoetes nuttallii* A. Br.
- *Isoetes occidentalis* Henderson
- *Isoetes truncata* (A. A. Eaton) Clute
- *Juncus supiniformis* Engelm.
- {*Lagarosiphon major*} Ridley
- *Lemna minor* L.
- *Lemna trisulca* L.
- *Lilaea scilloides* (Poir.) Hauman
- *Lilacopsis occidentalis* Coult. and Rose
- {*Limnobium laevigatum*} (Humb. and Bonpl.) Heine
- {*Limnobium spongia*} (Bosc) Steud.
- {*Limnophila sessiliflora*} Blume
- *Limosella aquatica* L.
- *Lobelia dortmanna* L.
- *Ludwigia palustris* (L.) Ell.
- *Lysimachia thyrsiflora* L.

- *Marsilea vestita* Hook. and Grev.
- *Megalodonta beckii* (Torr. ex Spreng.) Greene
- *Menyanthes trifoliata* L.
- *Montia fontana* L.
- {*Myriophyllum aquaticum*} (Vell.) Verd.
- *Myriophyllum farwellii* Morong
- {*Myriophyllum heterophyllum*} Michx.
- *Myriophyllum hippuroides* Nutt.
- *Myriophyllum pinnatum* (Walt.) B. S. P.
- *Myriophyllum quitense* H. B. K.
- *Myriophyllum sibiricum* Kom.
- *Myriophyllum spicatum* L.
- *Myriophyllum ussuriense* (Regel) Maxim.
- *Myriophyllum verticillatum* L.
- *Najas flexilis* (Willd.) R. and S.
- *Nasturtium officinale* R. Br. in W. Ait
- *Nitella acuminata* A. Br.
- *Nitella clavata* Kutz.
- *Nitella flexilis* (L.) Ag.
- *Nitella furcata* (Roxb.) Ag.
- *Nitella gracilis* (Sm.) Ag.
- *Nitella tenuissima* (Desv.) Kutz.
- *Nuphar polysepalum* Engelm.
- *Nuphar variegatum* Engelm. ex Durand
- *Nymphaea alba* L.
- *Nymphaea leibergii* Morong
- *Nymphaea mexicana* Zucc.
- *Nymphaea odorata* Ait.
- *Nymphaea tetragona* Georgi
- {*Nymphoides aquatica*} (Gmelin) O. Kuntze
- {*Nymphoides cordatum*} (Ell.) Fernald
- {*Nymphoides peltata*} (Gmelin) O. Kuntze
- *Phyllospadix scouleri* Hooker

- *Phyllospadix serrulatus* Rupr.
- *Phyllospadix torreyi* S. Wats.
- {*Pilularia americana*} R. Br.
- {*Pistia stratiotes*} L.
- *Polygonum amphibium* L.
- *Polygonum hyropiper* L.
- *Polygonum hydropiperoides* Michx.
- *Polygonum lapathifolium* L.
- {*Pontederia cordata*} Lour.
- *Potamogeton alpinus* Balbis
- *Potamogeton amplifolius* Tuckerman
- *Potamogeton berchtoldii* Fieb. in Bercht.
- *Potamogeton crispus* L.
- *Potamogeton epihydrus* Raf.
- *Potamogeton filiformis* Pers.
- *Potamogeton foliosus* Raf.
- *Potamogeton friesii* Rupr.
- *Potamogeton gramineus* L.
- *Potamogeton illinoensis* Morong
- *Potamogeton natans* L.
- *Potamogeton nodosus* Poir.
- *Potamogeton oakesianus* Robbins
- *Potamogeton obtusifolius* Mertens and Koch
- *Potamogeton pectinatus* L.
- *Potamogeton perfoliatus* L.
- *Potamogeton praelongus* Wulf.
- *Potamogeton pusillus* L.
- *Potamogeton richardsonii* (A. Benn.) Rydb.
- *Potamogeton robbinsii* Oakes
- *Potamogeton strictifolius* Bennett
- *Potamogeton vaginatus* Turcz.
- *Potamogeton zosteriformis* Fern.
- *Ranunculus aquatilis* L.

- *Ranunculus circinatus* Sibth.
- *Ranunculus cymbalaria* Pursh
- *Ranunculus flabellaris* Raf.
- *Ranunculus flammula* L.
- *Ranunculus gmelinii* DC.
- *Ranunculus hyperboreus* Rottb.
- *Ranunculus lobbii* (Hiern) A. Gray
- *Ranunculus sceleratus* L.
- *Riccia fluitans* L.
- *Ricciocarpus natans* (L.) Corda
- *Ruppia maritima* L.
- *Sagittaria cuneata* Sheld.
- *Sagittaria latifolia* Willd.
- *Salvinia* Seguier
- *Scheuchzeria palustris* L.
- *Scirpus lacustris* L.
- *Scirpus subterminalis* Torr.
- *Sium sauve* Walt.
- *Sparganium angustifolium* Mich.
- *Sparganium emersum* Rehm.
- *Sparganium eurycarpum* Engelm.
- *Sparganium fluctuans* (Morong) Robbins.
- *Sparganium glomeratum* Laest. ex Beurl.
- *Sparganium hyperboreum* Laest. ex Beurl.
- *Sparganium natans* L.
- *Spirodela polyrhiza* (L.) Schleid.
- *Subularia aquatica* L.
- *Tolypella intricata* (Trent.) Leonh.
- {*Trapa natans*} L.
- *Utricularia gibba* L.
- *Utricularia intermedia* Hayne
- *Utricularia minor* L.
- *Utricularia vulgaris* L.

- *Vallisneria americana* Michx.
- *Veronica anagallis-aquatica* L.
- *Veronica beccabunga* L.
- *Veronica catenata* Pennel
- *Veronica scutellata* L.
- *Wolffia borealis* (Engelm. ex Hegelmaier) Landolt and Wildi
- *Wolffia columbiana* Karst.
- *Wolffiella floridana* (J. D. Smith) Thompson
- *Zannichellia palustris* L.
- *Zostera japonica* Ascher and Grabn.
- *Zostera marina* L.

General Key to the Aquatic Plants of British Columbia

- plants of lakes, ponds, rivers and other permanent freshwater habitats, all herbaceous, rooted emergents, floating or submersed.
- plants all floating freely on the surface of the water, or just under the surface, not rooted or attached, except sometimes when stranded. General Key: Aquatic Plants of British Columbia: Part 1.
- plants submersed or emergent, generally rooted or attached, not freely floating on or at the surface.
- plants fully submersed, leaves may float on the surface and flowers may be emergent, stems and petioles remain on or under the water.
- plants with at least some, and often all, the underwater leaves finely dissected, and with at most a few floating leaves. General Key: Aquatic Plants of British Columbia: Part 2.
- plants lacking finely dissected submersed leaves, leaves may all float on the surface, may all be fully submersed or be some combination of both. General Key: Aquatic Plants of British Columbia: Part 3.
- plants rooted in the sediment but emergent, much, if not most or all, of the stem is emergent for most or all of the year.

- leaves cauline, more or less evenly distributed along the stem.
- leaves opposite or whorled, in groups, or clusters of several leaves. General Key: Aquatic Plants of British Columbia: Part 4.
- leaves alternate, only one at a node. General Key: Aquatic Plants of British Columbia: Part 5.
- leaves in basal clusters or bunches, not evenly distributed along an elongate stem. General Key: Aquatic Plants of British Columbia: Part 6.
- plants of marine or brackish lagoon habitats. General Key: Aquatic Plants of British Columbia: Part 7.

Part 1: Floating Plants

A microscope is needed for positive identification of *Azolla* species.

- Floating on the surface of the water.
- Large plants with distinct roots, stems and leaves; may have an obvious swollen part used as a floatation organ.
- most leaves are erect and emergent, in compact rosettes or distributed along the stem.
- petioles inflated to serve as floats, leaf bases cordate or abruptly narrowed to the petiole, flowers large, showy and purple, leaves glabrous and margins entire, rounded in outline, parallel veined. - {*Eichhornia crassipes*}
- leaves sessile or sub-sessile, no inflated petiole, leaves pubescent.
- plants with all leaves in a compact rosette arising from a common point, leaves obovate and apically notched, flowers in short inflorescences, shorter than the greyish-green and densely pubescent large leaves, flowering plants. - {*Pistia stratiotes*}
- plants with small hairy leaves distributed along the stems, ferns. - {*Salvinia*}
- leaves lie flat on the surface of the water, leaves long petiolate, arising from a common point.
- leaf bases cuneate to truncate, flowers short-petiolate and inconspicuous, leaves rhombic and angular, apical

margins at least toothed, leaves not thick and spongy but the petiole has a small swollen section, at the leaf junction, for a float. - {*Trapa natans*}

- leaves round to heart-shaped with a basal notch, leaf blades spongy, petioles without a swollen section. - {*Limnobium*}
- leaves with a thick spongy layer, convex on the dorsal surface, short petiole and closely clustered unmarked leaves. - {*Limnobium laevigatum*}
- leaves with a thin spongy layer, flat on the dorsal surface, petioles longer than the remote leaf blades, leaves with red-brown markings on the dorsal surface. - {*Limnobium spongia*}
- Small thalloid plants with roots below, closely appressed to the surface of the water.
- thalli not repeatedly and regularly forked, elongate to orbicular, flat or globose, flowering plants (duckweeds) and ferns (*Azolla*).
- ferns (*Azolla*), stem branched above every third leaf, leaves minute, alternate, sessile, bi-lobed, one lobe submersed as a float, the other emergent and colonized by blue-green algae. - *Azolla*
- glochidia with several (usually 3 or more) septa, leaves at least 0.7 mm long, plant 1 to 3 cm in diameter, submersed leaf-lobes much larger than the upper lobes, glochidia unbranched. - *Azolla mexicana*
- glochidia with 1 (2) or no septa.
- glochidia without any septa, leaves papillose, oblong to ovate, about 1 mm long, megaspores coarsely roughened, plants 1 to 10 cm long, branches numerous and open, lower submersed leaf-lobe about as large as the upper lobe, glochidia unbranched. - *Azolla filiculoides*
- glochidia with 1 or 2 apical septa, leaves nearly smooth, sub-orbicular, about 0.5 mm long, megaspores finely roughened, plants usually less than 3 cm long, few branches and crowded, lower submersed leaf-lobe glabrous, larger and paler than the upper lobe, glochidia may be branched. - *Azolla caroliniana*

- flowering plants (duckweeds), no elongate stem, leaves solitary or in clusters, not bi-lobed, no leaves specially adapted as floats, no algae colonies. - Lemnaceae
- one or more roots and nerves on each thallus (frond).
- each thallus bears two or more clustered roots from the base and 4 to 12 nerves. - *Spirodela polyrhiza*
- each thallus bears one root at the base and 1 to 3 nerves. - *Lemna*
- fronds oblong to lanceolate, 6 to 12 mm long and connected in small groups by stalks of the same length, matted, generally submersed, colonies. - *Lemna trisulca*
- fronds oval to round, less than 6 mm long, no stalks, solitary or in small attached groups, floating on the surface, usually 1 nerved, 2 to 4 mm long. - *Lemna minor*
- no roots or nerves on the thalli.
- thalli sub-globose to oval, up to 1 mm long. - >b>*Wolffia*
- plants floating just below the surface, sub-globose, upper surface rounded, green, not puncticulate, 0.5 to 1.0 mm long. - *Wolffia columbiana*
- plants floating on the surface, ellipsoidal, upper surface flattish, white- or brown-punctilate, 0.5 to 1.2 mm long and about 1/2 as wide. - *Wolffia borealis*
- thalli long and narrow, sickle-shaped, several mm long but very narrow, flattened, submersed except at the base and aggregated into clusters. - {*Wolffiella floridana*}
- thalli repeatedly and regularly forked, bryophytes (liverworts).
- thalli broad, 2 to 3 times forked, tinged with purple, and fringed with tongue-shaped toothed scales. - *Ricciocarpus natans*
- thalli narrow and elongate, repeatedly dichotomously forked, not purple tinged nor fringed with scales. - *Riccia fluitans*
- Floating beneath the surface of the water.
- Found just under the surface tension of the water or free in the mid-water zone; often found tangled in other submersed or marginal vegetation.

- plants with roots, stems and dissected leaves.
- plants with bladders on the alternate leaves. - *Utricularia*
- leaves divided into ultimately terete or threadlike segments.
- leaves divided into fewer than 5 final threadlike segments, leaf margins glabrous, bladders scarce on a small delicate plant usually floating at the surface or entangled in other rooted plants. - *Utricularia gibba*
- leaves 'pinnatifid', more than 20 terete final segments, hairy leaf margins, many bladders on the ordinary leaves, a robust plant to several meters long, usually lying on the sediment surface. - *Utricularia vulgaris*
- leaves divided di- or tri-chotomously into ultimately flattened segments, ordinary leaves generally with a few bladders, leaf margins glabrous, the terminal leaf segments are acuminate. - *Utricularia minor*
- plants without bladders on the whorled leaves. - *Ceratophyllum*
- leaf segments sub-capillary, mostly entire, delicate and light green, in deeper water and not surfacing, achene with 3 to 5 lateral spines on each side, not a 'weedy' species. - *Ceratophyllum echinatum*
- leaf segments capillary to linear and flattened, serrate to coarsely toothed, plant usually coarse and robust, dark green to almost black, usually surfacing, achene without lateral spines, 2 basal spines and 1 terminal spine only, a very 'weedy' species in eutrophic waters. - *Ceratophyllum demersum*
- plants reduced to thalli with no stems or leaves, roots may be present. - Lemnaceae
- one or more roots and nerves on each thallus, fronds oblong to lanceolate, 6 to 12 mm long and connected in small groups by stalks of the same length, matted, generally submersed, colonies. - *Lemna trisulca*
- no roots or nerves on the thalli.
- thalli sub-globose to oval, up to 1 mm long. - *Wolffia*

- plants floating just below the surface, sub-globose, upper surface rounded, green, not puncticulate, 0.5 to 1.0 mm long. - *Wolffia columbiana*
- plants floating on the surface, ellipsoidal, upper surface flattish, white-or brown-punctilate, 0.5 to 1.2 mm long and about 1/2 as wide. - *Wolffia borealis*
- thalli long and narrow, sickle-shaped, several mm long but very narrow, flattened, submersed except at the base and aggregated into clusters. - {*Wolffiella floridana*}
- Found lying on or just above the sediment surface.
- plants reduced to thalli with no stems or leaves, 1 root, fronds oblong to lanceolate, 6 to 12 mm long and connected in small groups by stalks of the same length, matted, generally submersed, colonies. - *Lemna trisulca*
- plants with roots, stems and dissected leaves bearing bladders. - *Utricularia*
- leaves divided into ultimately terete or threadlike segments.
- leaves divided into fewer than 5 final threadlike segments, leaf margins glabrous, bladders scarce on a small delicate plant usually floating near the surface or entangled in other rooted plants. - *Utricularia gibba*
- leaves 'pinnatifid', more than 20 terete final segments, hairy leaf margins, many bladders on the ordinary leaves, a robust plant to several meters long, usually lying on the sediment surface. - *Utricularia vulgaris*
- leaves divided di- or tri-chotomously into ultimately flattened segments, ordinary leaves generally with a few bladders, leaf margins glabrous, the terminal leaf segments are acuminate. - *Utricularia minor*

Part 2: Plants With Finely Dissected Submersed Leaves

- Submersed leaves pinnately divided, 1 to 5 cm long.
- submersed leaves bi-pinnate or simply pinnate with at least the lowest segments forked, one fork pinnate and the other simple. (aquarium plant) - {*Limnophila sessiliflora*}

- submersed leaves simply pinnate (native and introduced plants widely used in aquaria and outdoor pools). - *Myriophyllum*
- flowers in the axils of cauline submersed leaves.
- leaves in whorls of 3 to 4 or scattered, fewer than 10 leaf segments on each side of the rachis, monoecious, 4 stamens, plants often reddish, fully submersed. - *Myriophyllum farwellii*
- leaves in whorls of 4 to 6, more than 10 leaf segments on each side of the rachis, dioecious, 8 stamens, plants pale yellowish-green, the apical portion of the stem often sprawled over the surface of the water or on the adjacent shore. - {*Myriophyllum aquaticum*}
- flowers in the axils of bracts on emergent, terminal spikes.
- four stamens, 4 to 6 leaves per whorl, bracts conspicuous.
- floral bracts delicate, deeply incised to serrate, spike short and delicate, known only from sloughs in the Fraser Valley. - *Myriophyllum hippuroides* and *Myriophyllum pinnatum*
- floral bracts ovate and toothed, spike long, robust and inflated introduced in several park and garden ponds of the south-west. - {*Myriophyllum heterophyllum*}
- eight stamens, 3 to 5 leaves per whorl, floral bracts various.
- floral bracts smaller than the flowers, inconspicuous, and nearly entire (the lowest few may be larger and pinnate but the upper ones are small), leaf whorls in the central portion of the stem are over 1 cm apart and not crowded, monoecious.
- no turions, rhizomatous, 10 to 16 leaf divisions less than 2 mm apart, leaves make right or obtuse angles with the stem, leaf tips 'squared', all leaf segments straight and all of nearly the same length. - *Myriophyllum spicatum*
- turions, not rhizomatous, 6 to 12 leaf segments over 2.5 mm apart, leaves make acute angles with the stem, leaf tips 'acute', basal leaf segments curved and much longer

than the apical ones. - *Myriophyllum sibiricum* [*Myriophyllum exalbescens*]

- floral bracts usually longer than the flowers and rarely entire, leaf whorl spacing varies, monoecious or dioecious.
- dioecious, flowering plants found only on exposed mud banks when water levels drop in summer, female bracts and leaves entire to scarcely and irregularly divided, male bracts and leaves entire to pectinate-pinnate, submersed leaves scattered and irregular. - *Myriophyllum ussuriense*
- dioecious or monoecious, flowering plants found in water, bracts large and conspicuous, variable.
- floral bracts pinnate to pectinate, light green, leaves often crowded on the stem (short internodes) and delicate, usually with more than 10 leaf divisions. - *Myriophyllum verticillatum*
- floral bracts pectinately parted below becoming dentate in the middle and almost entire above, reddish, leaves well spaced on the stem (long internodes) and robust, leaves generally with fewer than10 leaf divisions. - *Myriophyllum quitense*
- submersed leaves palmately or di- to tri-chotomously divided but not pinnate, length is variable.
- leaves sessile, in whorls of 5 to 12, up to 2 cm long, toothed or serrate, plants are fully submersed and have no roots. - *Ceratophyllum*
- leaf segments sub-capillary, mostly entire, delicate and light green, deeper water and not surfacing, achene with 3 to 5 lateral spines on each side. - *Ceratophyllum echinatum*
- leaf segments capillary to linear and flattened, serrate to coarsely toothed, usually coarse and robust, dark green to almost black, usually surfacing, achene without lateral spines, 2 basal spines and 1 terminal spine only. - *Ceratophyllum demersum*
- leaves petiolate, in whorls of 3, decussate, entire, plants rooted. - {*Cabomba caroliniana*}

- submersed leaves alternate or opposite, up to 8 cm long, toothed or entire, plants submersed or emergent and with or without roots.
- submersed leaves opposite, 2 to 6 cm long, many times divided into ultimately filiform segments.
- leaves 2 to 4 cm long, sessile, many times dichotomously divided, the leaf segments remain filiform. - *Megalodonta beckii*
- leaves up to 6 cm long, petiolate, many times palmately divided, leaflet tips are flattened and spathulate. - {*Cabomba caroliniana*}
- submersed leaves alternate, 0.3 to 8 cm long, few to many times divided into flat filiform segments or threadlike.
- leaves with bladders, sessile, petiolar base of leaves not swollen. - *Utricularia*
- leaves divided into ultimately terete or threadlike segments.
- submersed leaves in whorls.
- leaves divided into fewer than 5 final threadlike segments, leaf margins glabrous, bladders scarce on a small delicate plant usually floating at the surface or entangled in other rooted plants. - *Utricularia gibba*
- leaves 'pinnatifid', more than 20 terete final segments, hairy leaf margins, many bladders on the ordinary leaves, a robust plant up to several meters long and usually lying on the sediment surface. - *Utricularia vulgaris*
- leaves divided di- or tri-chotomously into ultimately flattened segments.
- ordinary leaves generally with a few bladders, leaf margins glabrous, the terminal leaf segments are acuminate. - *Utricularia minor*
- ordinary leaves rarely with bladders, bladders are found on separate subterranean branches, leaf margins hairy, the ultimate leaf segment tips are awned. - *Utricularia intermedia*

- no bladders on the sessile or petiolate leaves.
- a swollen float at the top of the petiole next to the blade in the floating leaves which are rhomboid, flowers white, perianth 4-merous, fruit a unique, long-stalked nut. - {*Trapa natans*}.
- petiolar base of leaves is swollen, flowers white or yellow, fruit of 1-seeded nutlets, perianth 5-merous, emergent or floating leaves various but not rhomboid. - *Ranunculus*
- flowers white, submersed leaves dissected into ultimately filiform segments.
- Two to 7 glabrous, beakless achenes, plants and receptacle glabrous, pedicels in the axils of ternately lobed floating leaves, leaves 2 to 3 times divided into 8 to 12 segments. - *Ranunculus lobbii*
- Ten to 80 glabrous or hirsute achenes, plants and receptacles hirsute.
- Thirty to 80 achenes, flowers may have yellow bases, leaf blades sessile on the stipular base. - *Ranunculus circinatus*
- Ten to 25 achenes, flowers all white, leaf blades petiolate. - *Ranunculus aquatilis*
- flowers yellow, submersed leaves, if present, simple to ternately divided, leaves all or in part lobed, parted or ternately dissected to filiform segments, achenes pubescent.
- annuals, erect, not nodally rooting, achene beaks inconspicuous. - *Ranunculus sceleratus*
- perennials, floating or reclining, nodally rooting, achene beaks conspicuous.
- leaves deeply 3 parted with narrow lobes distally acute, achenes not corky margined, achene beaks 1/4 the length of the achenes. - *Ranunculus gmelinii*
- leaves, at least the submersed ones, 3 to 5 times ternately dissected into filiform segments less than 2 mm wide, achenes corky margined. - *Ranunculus flabellaris*

Part 3: Submersed Plants Without Finely Dissected Leaves

Section 1: Non-flowering Plants (Algae, Ferns and Quillworts- *Isoetes*):

A compound or dissecting microscope is required to positively identify species in *Isoetes*; magnification over 10x is needed to examine the megaspore surface.

- algae, plant body composed of 1-celled internodes, which may be covered with a multi-cellular cortex, and with branchlets at the nodes, no leaves, fully submersed, often foul smelling and encrusted with marl, spores reddish in spherical, axillary clusters. - Characeae
- branchlets simple, not further branched or segmented, coronula 5-celled. - *Chara*
- stem not corticate, stipulodes in 1 tier, and alternating with the branchlets, branchlets tipped with a corona, no spine cells. - *Chara braunii*
- stems, and at least basal branchlet segment, corticate.
- spine cells in fascicles, axial cortex haplostichous, 1 corticate, stipulodes in 2 tiers, spiny looking, found in saline or brackish waters. - *Chara canescens*
- spine cells solitary or geminate, axial cortex diplostichous or triplostichous, 2 or 3 corticate.
- axial cortex diplostichous, stipulodes in 2 tiers, spine cells absent or small, rarely geminate, quite variable and widespread in hard-water lakes, smooth but marl encrusted. - *Chara vulgaris*
- axial cortex triplostichous, stipulodes usually in 2 tiers but may be in 1 tier, obscure or absent, spine cells usually absent or obscure but may be clustered, variable and widespread in hard-water and soft-water lakes, smooth appearance. - *Chara globularis*
- branchlets compound, re-branched and/or segmented.
- branchlets simple but segmented like a string of sausages, end cells reduced and acute, heads coarse and like a bird's nest. - *Tolypella intricata*
- branchlets compound, re-branched, may be segmented or apparently so. - *Nitella*

- heteroclemous, 2 kinds of branchlets in a whorl, 1-forked alternating with 1-celled, compact, branchlets swollen. - *Nitella clavata*
- homeoclemous, all branchlets in a whorl the same, generally 1-forked.
- fertile heads small and densely compact, no mucous, generally no terminal dactyls on the sterile branchlets. - *Nitella acuminata*
- fertile heads absent or loose and not compact.
- dactyl apices not long acuminate, may be acute, blunt or apiculate, (a small variety may have 2-forked branchlets). - *Nitella flexilis*
- dactyl apices are long acuminate, gametangia on branchlets, generally not marl covered. - *Nitella acuminata*
- the end segment of the branchlets is 2-or-more-celled, the end cells are very much smaller than the penultimate cell, branchlets furcate, heads terminal, branchlets homeoclemous.
- robust plants, over 15 cm high, axes over 600 microns in diameter, dactyls may be small. - *Nitella furcata*
- smaller plants, seldom over 15 cm tall, axes 200 to 450 microns in diameter, dactyls uniform.
- lowest branchlet node fertile, dactyls 2- to 3-celled, heads may have mucous, oospore membrane granular or felt-like. - *Nitella gracilis*
- lowest branchlet node sterile, dactyls 2-celled, heads without mucous, oospore membrane reticulate. - *Nitella tenuissima*
- Quillworts or ferns, plant body composed of a multicellular stem with leaves.
- leaves circinate in the bud ('fiddleheads'), ferns with hairy sporocarps at the bases of the leaves, leaves arise in clusters from the nodes of a creeping, horizontal stem. - Marsiliaceae
- leaves with a slender petiole and a 4-foliate blade (like a 4-leaf clover). - *Marsilea vestita*

- leaves are filiform, without an expanded blade. - *Pilularia americana*
- leaves not circinate, quillworts, no hairy sporocarps, sessile leaves apically filiform with a wider, clasping base, arising in a tight spiral from the apex of a short, compact 2- or 3-lobed corm-like structure, sporangia on the inner face of the leaf base. - *Isoetes*
- corms 3-lobed, plants primarily terrestrial, grow in grassy, ephemeral pools which are wet in winter and dry in summer. - *Isoetes nuttallii*
- corms 2-lobed, plants primarily of lake bottoms but sometimes exposed in summer due to water level fluctuations.
- megaspores spiny.
- megaspores vary greatly in size and are often aborted, spines blunt and dense. (this is probably a hybrid of *Isoetes echinospora* and *Isoetes maritima*). - *Isoetes truncata*
- megaspores uniform in size, spines sharp, not crowded.
- plants flaccid, megaspore spines elongate with pointed tips, uniform in size, microspores smooth. - *Isoetes echinospora*
- plants rigid, megaspore spines stubby and blunt, shorter and denser along the ridges, microspores spiny. - *Isoetes maritima*
- megaspores smooth or with ridges but not spiny.
- lowermost leaves in 2 ranks, megaspores over 0.5 mm in diameter, microspores rugose and over 36μm long. - *Isoetes occidentalis*
- lowermost leaves spirally arranged, megaspores less than 0.5 mm in diameter, microspore spiny and less than 30μm long.
- base of the leaves blackened, hyaline wing margins of the leaves extending 1 to 5 cm above the sporangium, sporangia brown spotted. - *Isoetes howellii*
- base of the leaves green, hyaline wing margins of the leaves not extending more than 1 cm above the

sporangium, sporangia without colour. - *Isoetes bolanderi*

- the end segment of the branchlets is 1-celled.

Section 2: Flowering Plants (Angiosperms)

In *Nuphar* the large, showy, yellow, perianth members are sepals, the petals are smaller than the stamens and inconspicuous.

Mature fruits are generally needed for positive identification of *Callitriche.*

- leaves peltate, petiole attached inside the margin, and floating on the surface.
- alternate leaves arise near the apex of a vertical floating stem, leaves and plant covered in a gelatinous sheath, leaves entire, oval and reddish, flowers solitary, reddish, on long peduncles from the upper axils, deep water plant, fruit a follicle. - *Brasenia schreberi*
- leaves solitary from the nodes of a horizontal creeping rhizome, no gelatinous sheath, leaves rounded, margins not entire, leaves green, flowers in umbels on a long peduncles from the nodes, shallow water plants, fruit a schizocarp. - *Hydrocotyle verticillata*
- leaves not peltate, petiole attached on the leaf margin, there may be a deep sinus with leaves apparently peltate in some Nymphaea.
- leaves arise in basal clusters, (plants scapose) or from relatively short, usually apical, sections of rhizomes, but not cauline and spread along the stem.
- leaves all fully submersed and much longer than wide.
- leaves with long petioles and lanceolate blades, solitary flowers on stalks shorter than the leaves, arising in a cluster among the leaves. - *Limosella aquatica*
- leaves sessile, flowers on stalks, exceeding or equalling the leaves.
- leaves very long and tape-like, flowers solitary on very long peduncles which reach the surface and coil up like a spring after fertilization and draw the fruit underwater. - *Vallisneria americana*

- leaves shorter than the flower stalks, flowers on few-flowered racemes, no coiled peduncle.
- leaves linear and sub-terete, not fleshy, several few flowered racemes, roots not white and conspicuous. - *Subularia aquatica*
- leaves thick and fleshy, terete or angular, 1 few-flowered raceme, plants with a dense cluster of white roots. - *Lobelia dortmanna*
- leaves, or some at least, with long petioles and floating on the surface, blades nearly round and with a deep sinus.
- tubers or slender rhizomes, 5 petals and sepals united at the base, leaves small, under 15 cm in diameter, flowers white or yellow. - *Nymphoides*
- flowers yellow, axillary, no cluster of roots on the petiole, leaves primarily from branching stems. - *Nymphoides peltata*
- flowers white, in clusters on the petioles, a cluster of roots on the petiole, leaves mostly basal with long slender petioles.
- leaf blades ovate, mostly under 5 cm long, seeds smooth, leaves not dark-punctate beneath. - *Nymphoides cordatum*
- leaf blades orbicular, 8 to 15 cm in diameter, seeds glandular, warty, leaves dark-punctate or pitted beneath. - *Nymphoides aquaticum*
- stout or massive rhizomes, numerous petals, 5 to 12 sepals, leaves large, some over 15 cm in diameter, flowers white, yellow, red, or pink.
- flowers yellow, 5 to 12 sepals, outer ones green, inner ones yellow tinged with red or green, no arils on the seeds, superior ovary, petals erect and smaller than the sepals or stamens, 1 discoid stigma. - *Nuphar*
- Six to 8 (usually 6) sepals, 2.5 to 3.5 cm long, yellow stamens. - *Nuphar variegatum*
- Eight to 17 (usually 9) sepals, (3) 3.5 to 6 cm long, reddish stamens. - *Nuphar polysepalum*

- flowers white, red or (yellow), sepals 4, green or streaked with red, seeds arillate, semi-inferior ovary, petals spreading, many stigmas. - *Nymphaea*
- flowers yellow, 6 to 13 cm in diameter, leaves dark green and blotched above, brownish with black dots below, plant tuberous. - *Nymphaea mexicana*
- flowers white pink or red, leaves not blotched.
- small plant with leaves up to 8 cm wide and flowers up to 5 cm wide, 7 to 15 petals, 6 to 9 stigmas, flowers white.
- carpellary appendages 3mm or longer, flaccid and purplish. - *Nymphaea tetragona*
- carpellary appendages 1.5mm or shorter, stiff and greenish. - *Nymphaea leibergii*
- larger plants with leaves over 8 cm wide and flowers over 5 cm wide, petals and stigmas numerous, flowers white, pink, or rose.
- flowers white or tinged with pink, 20 to 32 petals, leaves scattered on the rhizome, flowers strongly scented, leaves usually purplish beneath. - *Nymphaea odorata*
- flowers white, 12 to 24 petals, leaves crowded on the rhizome, flowers not strongly scented, leaves greenish beneath. - *Nymphaea alba*
- leaves cauline and more or less evenly distributed along the stem.
- horizontal, creeping rhizomes or stolons with leaves arising at the nodes, often rooted at every node.
- leaves bladeless and sessile, clustered, terete, septate, glabrous, flowers in short stalked umbels, shorter than the leaves and arising among the leaves at the nodes, stolons creeping. - *Lilaeopsis occidentalis*
- leaves with a blade and a petiole.
- solitary deeply lobed circular leaves at each node, small inconspicuous flowers in short-stalked, recurved umbels, one umbel at each node, stolons tending to creep. - *Hydrocotyle ranunculoides*

- leaves lobed and solitary to lanceolate and several, flowers solitary in the leaf axils of upper nodes, conspicuous and yellow or white, stolons creeping or arched between nodes. - Ranunculaceae
- several lanceolate, entire leaves at each node, stolons often arched between nodes, (apical leaves may be sessile), flower stalks longer than the apical leaves but shorter than the long-petiolate basal leaves. - *Ranunculus flammula*
- leaves solitary and deeply lobed at most nodes, stolons mostly flat, flower stalks about as long as the leaves.
- flowers yellow, sepals and petals, leaf margins entire, leaves 3 -lobed, 1 to 1.5 cm wide, fruit an ovoid head of achenes. - *Ranunculus hyperboreus*
- flowers white, sepals but no petals, leaf margins dentate, leaves reniform, 2 to 4 cm wide, fruit a cluster of follicles. - *Caltha natans*
- more or less erect or sprawling stems, not buried or rooted at every node.
- leaves opposite or whorled, several per node (in juvenile *Najas* with short internodes this may be difficult to determine).
- leaves opposite, 2 at a node.
- leaves capillary with membranous, sheathing stipules, stigma peltate, fruits clusters of short-stalked, banana-shaped, conspicuous achenes in the leaf axils, 1 stamen with a filament longer than the pistils, stems filiform and freely branched. - *Zannichellia palustris*
- leaves at least linear and sometimes ovate, no stipules, stigmas not peltate, fruit not a cluster of banana-shaped achenes, stamens various, stems not filiform and branched or unbranched.
- leaf with a clasping, short, broad base abruptly narrowing to long-attenuate, toothed on the shoulder, fruits solitary, fusiform, smooth and shining, achenes in the leaf axils, stamen almost sessile, plants branched and bushy apically and fragile, often fragmenting. - *Najas flexilis*

- leaves linear to obovate, not with a sheathing base and not abruptly narrowing, fruits axillary and bi-lobed or several-seeded capsules, stamens solitary with a long filament or 9 on stalked flowers, stems often unbranched or with a few apical branches, not usually bushy.
- leaves linear to narrowly lanceolate, 1-nerved, usually paired basally and whorled apically, 9 stamens, long-stalked flowers, fruit of several-seeded capsules. - *Elodea*
- leaves rarely over 1.5 cm long.
- leaves 2 (1 to 4) mm wide, tapered abruptly to a blunt point, staminate flowers stalked and persistent. - *Elodea canadensis*
- leaves 1.5 (0.3 to 1.5) mm wide, tapered to a slender point, staminate flowers sessile and deciduous at anthesis. - *Elodea nuttallii*
- leaves (1.7) 2.0 to 2.6 cm long. - {*Elodea longivaginata*}
- leaves linear to obovate or spathulate, 1 to 3 (5) nerved, all opposite, sometimes linear basally and obovate and clustered apically, fruit of bi-lobed, often winged, sessile, axillary, naked achenes. - *Callitriche*
- fruit encircled by a conspicuous wing-like margin, leaf bases joined by winged ridges. - *Callitriche stagnalis*
- fruit not winged or with only a narrow wing at the tip, leaf bases various.
- leaves all linear, 1-nerved, light green, leaf bases not joined by a wing, floral bracts absent, common species. - *Callitriche hermaphroditica*
- leaves various, upper often ovate and 3-nerved, bases joined by a wing-like ridge, floral bracts present.
- carpel face markings in regular vertical lines, fruit slightly wing-margined at the top and longer than broad. - *Callitriche verna*
- carpel face markings scattered, fruit not winged and as long as broad.
- fruits widest above the middle (obovate) leaves bidentate, midvein barely thickened at the end,

emergent leaves may be over 5 mm wide, stems long. - *Callitriche heterophylla*

- fruits round or oblong, midvein thickened at the, tip and protruding, leaves never over 5 mm wide, plants short and slender. - *Callitriche anceps*
- leaves whorled, 3 or more at a node.
- leaves (8) 10 to 25 (50) mm long in whorls of (4) 6 to 12, 1 style, 1 stamen, fruit nut-like, flowers solitary and sessile in the axils, glabrous perennials from creeping rhizomes, the apical portion of the stem may be emergent or the plant may grow completely emergent on wet mud with much shorter leaves and shorter internodes. - *Hippuris vulgaris*
- leaves under 25 (40) mm long in whorls of (2) 3 to 12, 3 or more styles, 3, 6 or 9 stamens, fruit a few-seeded capsule, flowers axillary and sessile but on long, thin stalks, the hypanthium, which reaches the surface, permanently submersed. - Hydrocharitaceae
- leaf arrangement irregular, margins serrate or toothed, tip with 2 enlarged spines. - {*Lagarosiphon major*}
- leaves arranged in regular whorls.
- leaf margins serrate or toothed.
- leaf tip with 2 enlarged spines, no turions , 3 or more strongly recurved leaves up to 3 cm long per whorl. - {*Lagarosiphon major*}
- no enlarged spines on the leaf tip, turions present on the rhizome or on the stem, 3 to 8 (12) nearly straight leaves up to 2.5 cm long per whorl. - {*Hydrilla verticillata*}
- leaf margins entire.
- basal leaves in whorls of 3 but upper ones up to 6 per whorl and up to 4 cm long, 2 to 3 flowers in the staminate spathes, petals to 10 mm. - {*Egeria densa*}
- basal leaves in pairs, the upper ones usually in whorls of 3 and up to 2.5 cm long, flowers solitary in the staminate spathes, petals up to 5 mm long. - *Elodea*
- leaves in whorls of 3.

- leaves rarely over 1.5 cm long.
- leaves 2 (1 to 4) mm wide, tapered abruptly to a blunt point, staminate flowers stalked and persistent. - *Elodea canadensis*
- leaves 1.5 (0.3 to 1.5) mm wide, tapered to a slender point, staminate flowers sessile and deciduous at anthesis. - *Elodea nuttallii*
- leaves (1.7) 2.0 to 2.6 cm long. - {*Elodea longivaginata*}
- leaves in the upper and middle part of the stem in pairs, lower ones irregular or alternate, (1.7) 2.0 to 2.6 cm long. - (*Elodea longivaginata*)
- plants sprawling at or just under the surface, nodally rooting, flowers showy, rose in terminal spicate panicles or pale yellow and solitary from the upper axils.
- flowers rose in a compact, terminal panicle, leaves broadly lanceolate to elliptic, floating on the surface, petiolate, leathery. - *Polygonum amphibium*
- flowers pale yellow, solitary on long stalks, in the axils of apical leaves, submersed, long and narrow, delicate. - *Heteranthera dubia*
- plants mostly erect, not usually rooting at the nodes, flowers reduced and in whorls on pedunculate spikes, inconspicuous. - *Potamogeton*
- stipules forming a sheath around the stem partly below the base of the leaf blade and partly above, the leaf attached 1 cm or more above the node.
- leaves 3 to 4 mm wide, flat, rough on the edges, stiff, with a broad mid-vein and over 20 lateral veins. - *Potamogeton robbinsii*
- leaves less than 3 mm wide, cylindrical, smooth, soft, with only one inconspicuous vein.
- leaves acute and sharp pointed at the apex, fruits 2.5 to 4 mm long with a short beak, the surface of fruits without radial striation (at 10x). - *Potamogeton pectinatus*
- leaves obtuse and blunt at the apex, fruits 2 to 3 mm long, beakless, surface of the fruits with radial striation (at 10x).

- primary stems 0.5 to 1 mm in diameter, all leaves filiform, 0.2 to 0.5 mm wide with tight sheaths, spikes with 2 to 5 whorls of flowers, fruits 2 to 2.5 mm long. - *Potamogeton filiformis*
- primary stems 1 to 3 mm in diameter, lower main stem leaves with blades 1 to 2 mm wide and with loose inflated sheaths, upper branch leaves filiform, spikes with 5 to 12 whorls of flowers, fruits 2.5 to 2.8 mm long. - *Potamogeton vaginatus*
- stipules forming a sheath around the stem only above the base of the leaf blade, the leaf attached at the node.
- submersed leaves linear, ribbon-like or cylindrical, less than 5 mm wide, parallel margins.
- submersed leaves cylindrical, terete.
- floating leaves 2.5 to 6 cm wide, usually cordate, large ones with (18) 21 to 35 nerves, submersed leaves arise from the main stem, mature fruit 3.7 to 4.5 mm long including the beak, obscurely keeled, interlacunar bundles in several circles throughout the aerenchyma. - *Potamogeton natans*
- floating leaves 1 to 3 cm wide, usually rounded or cuneate at the base, largest ones with 11 to 19 nerves, submersed leaves arise from the branches of the main stem, mature fruit, including the beak, 3 to 3.4 mm long, interlacunar bundles develop only in the outer circle of the aerenchyma. - *Potamogeton oakesianus*
- submersed leaves flat.
- leaves alternate, 1 per node.
- submersed leaves with 9 to 35 veins. - *Potamogeton zosteriformis*
- submersed leaves with 1 to 7 veins.
- stipules strongly fibrous, becoming whitish, especially on the turions, base of turions strongly ribbed.
- leaves obtuse or rounded and slightly mucronate, not conspicuously 2 ranked,, blades thin, 1.5 to 3.5 mm wide, 5 to 7 veins, a narrow cellular-reticulate band along the midrib, turions fan-shaped, peduncle flattened. - *Potamogeton friesii*

- leaves gradually tapered into a sharp bristle tip, conspicuously 2 ranked, blades firm, 0.5 to 2.5 mm wide, convolute with 3 (5) veins, no cellular-reticulate band along the midrib, turions slender, peduncles terete. - *Potamogeton strictifolius*
- stipules delicate, not fibrous, greenish or brownish, base of the turions smooth.
- leaves (2) 3 to 4 mm wide, rounded at the apex, fruits (3) 3.5 to 4 mm long. - *Potamogeton obtusifolius*
- leaves 0.3 to 3 mm wide, acute to obtuse or mucronate, fruits 1.8 to 2.8 mm long
- fruits with a distinct dorsal keel, veins on the stipules evident as ridges running the full length of the stipule, glands at the base of the stipules either lacking or poorly developed. - *Potamogeton foliosus*
- fruits without a dorsal keel, veins on the stipules obscure and faint, glands at the base of the stipules usually well developed.
- stipules connate, fused along the stem, at least when young, mature fruits broadest above the middle, plants sparsely branched. - *Potamogeton pusillus*
- stipules convolute, mature fruits broadest below the middle, plants well branched. - *Potamogeton berchtoldii*
- submersed leaves lanceolate or ovate, over 5 mm wide.
- leaf margins serrate, beak of the fruit as long as the fruit body or longer. - *Potamogeton crispus*
- leaf margins entire, beak of the fruit much shorter than the fruit body.
- submersed leaves ribbon-like with parallel side, 5 to 10 mm wide, limp and flaccid, median band of lacunae several cells wide at least 1/4 the width of the blade, stems compressed. - *Potamogeton epihydrus*
- submersed leaves lanceolate to ovate with sides not parallel, either no median cellular-reticulate band or the band less than 1/4 the width of the blade, stems terete.
- submersed leaves sessile, cordate or rounded at the base and clasping the stem.

- leaves ovate-oblong, mostly 10 to 20 cm long, with a cucullate apex, fruits more than 4 mm long, stems with many interlacunar vascular bundles. - *Potamogeton praelongus*
- leaves rounded, ovate or elongate-ovate, 1 to 10 cm long, no cucullate apex, fruits under 3.5 mm long, no interlacunar vascular bundles.
- stipules coarse disintegrating into persistent whitish fibres, peduncles clavate, 1.5 to 25 cm long, fruits with a cavity in the endocarp loop. - *Potamogeton richardsonii*
- stipules delicate, lacking on mature specimens, peduncles not clavate, 1 to 9 cm long, fruits without a cavity in the endocarp loop. - *Potamogeton perfoliatus*
- submersed leaves petiolate or sessile, not clasping the stems.
- submersed leaves 2 to 5 cm wide, tapering into distinct petioles, stipules over 3 cm long.
- floating leaves cordate, submersed leaves folded and strongly falcate, more than 20 veins, interlacunar bundles well developed throughout. - *Potamogeton amplifolius*
- floating leaves rounded or cuneate at the base, submersed leaves flat and not falcate, with fewer than 20 veins, interlacunar bundles developed in only one circle or absent.
- petioles of submersed leaves shorter than 2 cm, interlacunar bundles well developed forming one circle, endodermis of U-cells. - *Potamogeton illinoensis*
- petioles of submersed leaves longer than 2 cm, interlacunar bundles absent, endodermis of O-cells. - *Potamogeton nodosus*
- submersed leaves mostly less than 2 cm wide, sessile, or with petioles under 0.5 cm long, stipules under 3 cm long
- plants with a reddish tinge, usually not branched, floating leaves, if present, not markedly different from the submersed leaves, no interlacunar bundles, endodermis of O-cells. - *Potamogeton alpinus*

- plants greenish, freely branched, floating leaves markedly different from the submersed leaves, interlacunar bundles well developed in one circle, endodermis of U-cells. - *Potamogeton gramineus*

Part 4: Emergents- Opposite or Whorled Cauline Leaves

Our aquatic *Equisetum* all share the following characteristics: stems annual, usually with whorls of branches, distinct sterile and fertile stems, cones blunt.

- leaves in whorls of 3 or more at a node, small and scale or bract-like, basally sheathing, erect stems with longitudinal grooves offset at each node, sporangia in terminal cones, non-flowering plants, horsetails. - *Equisetum*
- central cavity well over 1/2 the diameter of the stem, teeth may be deciduous, 10 to 40 ridges, teeth persistent, black but not hyaline margined, stomates in 1 row and not sunken. - *Equisetum fluviatile*
- central cavity less than 1/3 the diameter of the stem, teeth not deciduous, black and hyaline margined, 5 to 10 ridges. - *Equisetum palustre*
- leaves opposite, only 2 at a node, flowering plants.
- flowers sessile in the axils, 2 to 4-merous perianth, capsules.
- leaves under 1.5 cm long, spathulate, flowers (2) 3 merous, seeds curved and pitted. - *Elatine rubella*
- leaves 2 to 6 cm long, ovate to elliptic, no petals, flowers 4-merous, seeds glabrous. - *Ludwigia palustris*
- flowers on stalks in the axils.
- plants sprawling or creeping, if more or less erect under 1 dm tall, not woody or spongy at the base, flowers not in whorls in the axes or in congested racemes (*Lysimachia thyrsiflora*).
- flowers solitary in the axils.
- five sepals, 2-lobed corolla tube, 2 stamens and stigmas, fruit a capsule. - *Gratiola*
- there are 2 sepaloid bracts at the apex of the pedicel and thus apparently 7 sepals. - *Gratiola neglecta*

- pedicels without bracts and sepals thus evidently only 5. - *Gratiola ebracteata*
- flowers greenish, 4 sepals and petals, 4 stamens, fruit of 4 follicles. - *Crassula aquatica*
- flowers in open racemes, 2 sepals, 3 stamens, 3 stigmas, sprawling and creeping, rooting at the nodes, annual. - *Montia fontana*
- plants more or less erect to apically drooping, over 1 dm tall.
- flowers crowded in short, dense racemes axillary in the middle leaves, flowers 5 (6 to 7) merous. - *Lysimachia thyrsiflora*
- flowers in elongate, open racemes, 4 (5) sepals, corolla 4 lobed, 2 stamens, capitate stigma. - *Veronica*
- leaves all short-petiolate. - *Veronica beccabunga*
- leaves, at least the middle and upper ones, sessile and clasping.
- capsules much wider than high, notched at the tip,, leaves (three) 4 to 20 times as long as wide. - *Veronica scutellata*
- capsules not wider than high, scarcely, if at all notched at the tip,, leaves 1.5 to 5 times as long as wide.
- leaves 1.5 to 3 times as long as wide, fruiting pedicels ascending or upcurved, flowers blue or violet. - *Veronica* anagallis-aquatica
- leaves (2.5) 3 to 5 times as long as wide, fruiting pedicels spreading, flowers white to pink or pale blue. - *Veronica catenata*

Part 5: Emergents- Alternate Cauline Leaves

- leaves compound.
- flowers in umbels, (Apiaceae or Umbelliferae, base of the stem without transverse septa, roots not tuberous thickened, ribs of the fruit prominent, corky-thickened or covered in prickles, primary lateral veins of the leaflets not directed in any particular direction, plants erect, calyx teeth tiny or absent). - *Sium sauve*

- flowers in axillary racemes, (fruit a silique, plants prostrate and nodally rooting, plant used as a salad green, watercress, seeds biseriate). - *Nasturtium officinale*
- leaves simple.
- leaf margins lobed or toothed.
- leaves lobed, flowers yellow, fruit a head of achenes, inflorescence axillary. - *Ranunculus hyperboreus*
- leaves only toothed or crenulate, long-petiolate, rounded with a deep basal sinus, margin crenate, flowers solitary on long peduncles. - *Caltha*
- sepals yellow, no petals, plants erect or sprawling. - *Caltha palustris*
- sepals white, no petals, plants creeping or floating. - *Caltha natans*
- leaf margins entire or nearly so.
- flowers naked or subtended by chaffy or bristle-like bracts.
- flowers mostly perfect, if imperfect then in reduced spikes with the staminate often not uppermost, grasses or sedges
- leaves in 3 vertical ranks on an often triangular, solid or pithy, stem without swollen nodes, sheaths closed, flowers subtended by 1 (2) bract (s) and often with inner subtending scales or bristles, fruit a usually beaked achene, ovary sometimes enclosed in a sac-like covering, styles often 3. - Cyperaceae
- leaves in 2 vertical ranks on a usually terete, hollow stem with swollen nodes, sheaths commonly open but may be closed, flowers usually subtended by 2 bracts with 2 obscure inner subtending scales, ovary never enclosed in a sac, fruit a grain with the ovary wall adherent to the seed, rarely beaked, up to 2 styles, grasses. - Poaceae
- flowers mostly imperfect, several in globose to capitate clusters, each subtended by 3 to 5 chaffy bracts, fruit hard, beaked, conspicuous, staminate and pistillate in separate heads not grass-like (bur-reeds). - *Sparganium*

- stigmas over 2 mm long, the achenes truncate-pyriform and narrowed abruptly to the beak, inflorescence usually branched, staminate heads above the pistillate heads, large robust plants. - *Sparganium eurycarpum*
- stigma, achenes fusiform and tapering to the beak, inflorescence usually simple, staminate heads below the pistillate heads.
- achene beaks less than 1.5 mm long or beakless, staminate head solitary.
- achene beaks less than 0.5 mm long to absent, staminate head closely adjacent to the upper pistillate head, basal leaves (0.5) 1 to 3 (5) mm wide, opaque, often yellow. - *Sparganium hyperboreum*
- achene beaks 0.5 to 1.5 mm long, staminate head distant from the upper pistillate head, basal leaves (1.5) 2 to 6 (10) mm wide, translucent, usually dark green. - *Sparganium natans*
- achene beaks 1.5 to 5 mm long, there may be more than 1 staminate head.
- achene beaks conspicuously curved. - *Sparganium fluctuans*
- leaves parallel veined, floral parts in 3's (4's) but not 5's. - Monocotyledons
- achene beaks straight or slightly curved.
- staminate head usually solitary, closely adjacent to the upper pistillate head. - *Sparganium glomeratum*
- staminate heads (1) 3 to 8, distant from the upper pistillate head.
- pistillate heads (1) 2 to 3 (4), usually crowded and appearing as one elongated head terminating the inflorescence, achene beaks 1.5 to 2.0 (2.2) mm long. - *Sparganium angustifolium*
- pistillate heads (3) 4 to 7 (10), distant and distinct, achene beaks 2 to 4.5 (6) mm long. - *Sparganium emersum*
- flowers with a perianth, all alike or with distinct sepals and petals.

- flowers imperfect, several in globose to capitate clusters, each subtended by 3 to 5 chaffy bracts, fruit hard, beaked, conspicuous, (bur-reeds). - *Sparganium*
- flowers perfect or if imperfect (rare) then staminate and pistillate flowers not in separate capitate clusters and not subtended by 3 to 5 chaffy bracts, fruit various.
- inflorescence a terminal fleshy spike subtended by a large, usually whitish, showy spathe, flowers closely crowded and embedded, perianth and stamens in 4's, large red berries. - *Calla palustris*
- inflorescence not a fleshy spike with embedded flowers and a spathe, perianth and stamens usually other than 4.
- leaves in 3 vertical ranks on an often triangular, solid or pithy, stem without swollen nodes, sheaths closed, flowers subtended by 1 (2) bract (s) and often with inner subtending scales or bristles, fruit a usually beaked achene, ovary sometimes enclosed in a sac-like covering, styles often 3. - Cyperaceae
- leaves mostly not 3-ranked, stems usually terete, rarely grass-like, perianth segments usually in 2 series of 3 each.
- pistils more than 1, distinct or separating at maturity into 3 to 6 follicles, perianth inconspicuous, greenish, segments all alike, 3 to 6 carpels.
- flowers in a 3 to 12 flowered bracteate raceme, fruiting pedicels 12 to 25 mm long, follicles compressed, divergent, 1 or 2 seeded. - *Scheuchzeria palustris*
- flowers in more than 12 flowered, bracteate racemes or spikes, fruiting pedicels rarely up to 6 mm long, follicles neither compressed nor divergent, 1 seeded, no perianth, pistillate flowers axillary, perfect flowers in pedunculate spikes, long styles, fruit 3-lobed, an annual plant. - *Lilaea scilloides*
- 1 compound pistil, 1 or 3 celled, fruit not a follicle.
- plants grass-like, flowers inconspicuous, perianth greenish, purplish or brownish, segments all alike. - *Juncus supiniformis*

- plants not grass-like, flowers showy, blue or purple, 6 stamens, superior ovary, long-petiolate, sheathing, arrow-shaped, hastate, leaves, perianth fused, inflorescence subtended by 2 spathes, introduced. - *Pontederia cordata*
- leaves pinnately or palmately veined, flower parts mostly in 5's, sometimes 4's, rarely 3's. - Dicotyledons
- leaves long-petiolate, often longer than the blade which is lanceolate, plants creeping and nodally rooting, fruit a globose cluster of achenes, flowers yellow. - *Ranunculus flammula*
- leaves sessile or short-petiolate, plants usually more or less erect but if creeping or nodally rooting then the flowers blue in a lax, helicoid, raceme.
- leaves with generally lacerate, sheathing stipules, nodes swollen, 2 or 3 styles, (4 or 6) 8 stamens. - *Polygonum* (in part)
- perianths glandular-punctate, glands sessile, achenes brown, glandular and dull, perianth usually 4-lobed, stamens usually 4 or 6. - *Polygonum hydropiper*
- perianths not glandular-punctate, achenes dark brown or black, perianth 4 to 5 lobed, stamens 8.
- inflorescence ending in spicate, slender, interrupted racemes mostly over 3 cm long, nerves of the perianth segments not branched and recurved, stamens 8, perianth 5-lobed. - *Polygonum hydropiperoides*
- inflorescence of several short, thick, continuous racemes, rarely over 3 cm long, nerves of the perianth segments branched and recurved, perianth 4 to 5 lobed. - *Polygonum lapathifolium*
- leaves without lacerate, sheathing stipules, nodes not swollen, one style, 5 stamens, flowers in helicoid, sympodial false racemes (forget me not). - *Myosotis*
- corolla limb 2 to 5 mm wide, style shorter than the nutlets, not stoloniferous, often recumbent. - *Myosotis laxa*
- corolla limb 5 to 10 mm wide, style as long as or longer than the nutlets, stoloniferous, mostly erect and not creeping or recumbent. - *Myosotis scorpioides*

Part 6: Emergents- Basal or Terminal Leaf Clusters

In *Sagittaria* the usual non-achene key characters of bract length and shape and pedicel lengths have been found to be inconsistent and sufficiently variable even within one population as to be virtually useless as key characters. Many flowering, but not fruiting, collections are difficult to identify.

- leaves compound, flowering plants, with several leaves, leaves trifoliate, leaflets obovate and entire to undulate-dentate, petals white and densely scaly or hairy, flowers in racemes, fruit a capsule. - *Menyanthes trifoliata*
- leaves simple.
- non-flowering plants, spores, no flowers, quillworts, cluster of basally sheathing, narrow, terete leaves with spores in the axils of the expanded base. - *Isoetes nuttallii*
- flowering plants, with several leaves, flowers and seeds are formed, not spores, leaves not as above or with sporangia in the axils.
- leaves parallel veined, floral parts in 3's (4's) but not 5's, (*Calla* and *Pontederia* do not have parallel-veined leaves but have spathes). - Monocotyledons
- flowers naked or subtended by chaffy or bristle-like bracts.
- flowers mostly perfect, if imperfect then in reduced spikes with the staminate often not uppermost, grasses or sedges
- leaves in 3 vertical ranks on an often triangular, solid or pithy, stem without swollen nodes, sheaths closed, flowers subtended by 1 (2) bract(s) and often with inner subtending scales or bristles, fruit a usually beaked achene, ovary sometimes enclosed in a sac-like covering, styles often 3. - Cyperaceae
- leaves in 2 vertical ranks on a usually terete, hollow stem with swollen nodes, sheaths commonly open but may be closed, flowers usually subtended by 2 bracts with 2 obscure inner subtending scales, ovary never enclosed in a sac, fruit a grain with the ovary wall adherent to the seed, rarely beaked, up to 2 styles, grasses. - Poaceae

- flowers mostly imperfect, several in globose to capitate clusters, each subtended by three to 5 chaffy bracts, fruit hard, beaked, conspicuous, staminate and pistillate in separate heads, not grass-like, (bur-reeds). - *Sparganium*
- two (one) stigmas over 2 mm long, the achenes truncate-pyriform and narrowed abruptly to the beak, inflorescence usually branched, staminate heads above the pistillate heads, large robust plants. - *Sparganium eurycarpum*
- one stigma, achenes fusiform and tapering to the beak, inflorescence usually simple, staminate heads below the pistillate heads.
- achene beaks less than 1.5 mm long or beakless, staminate head solitary.
- achene beaks less than 0.5 mm long to absent, staminate head closely adjacent to the upper pistillate head, basal leaves (0.5) 1 to 3 (5) mm wide, opaque, often yellow. - *Sparganium hyperboreum*
- achene beaks 0.5 to 1.5 mm long, staminate head distant from the upper pistillate head, basal leaves (1.5) 2 to 6 (10) mm wide, translucent, usually dark green. - *Sparganium natans*
- achene beaks 1.5 to 5 mm long, there may be more than one staminate head.
- achene beaks conspicuously curved. - *Sparganium fluctuans*
- achene beaks straight or slightly curved.
- staminate head usually solitary, closely adjacent to the upper pistillate head. - *Sparganium glomeratum*
- staminate heads (1) 3 to 8, distant from the upper pistillate head.
- pistillate heads (1) 2 to 3 (4), usually crowded and appearing as one elongated head terminating the inflorescence, achene beaks 1.5 to 2.0 (2.2) mm long. - *Sparganium angustifolium*
- pistillate heads (3) 4 to 7 (10), distant and distinct, achene beaks 2 to 4.5 (6) mm long. - *Sparganium emersum*

- flowers with a perianth, all alike or with distinct sepals and petals.
- flowers imperfect, several in globose to capitate clusters, each subtended by 3 to 5 chaffy bracts, fruit hard, beaked, conspicuous, (bur-reeds). - *Sparganium*
- flowers perfect or if imperfect (rare) then staminate and pistillate flowers not in separate capitate clusters and not subtended by 3 to 5 chaffy bracts, fruit various.
- inflorescence a terminal fleshy spike subtended by a large, usually whitish, showy spathe, flowers closely crowded and embedded, perianth and stamens in 4's, large red berries. - *Calla palustris*
- inflorescence not a fleshy spike with embedded flowers and a spathe, perianth and stamens usually other than 4.
- leaves in 3 vertical ranks on an often triangular, solid or pithy, stem without swollen nodes, sheaths closed, flowers subtended by one (2) bract (s) and often with inner subtending scales or bristles, fruit a usually beaked achene, ovary sometimes enclosed in a sac-like covering, styles often 3. - Cyperaceae
- leaves mostly not 3-ranked, stems usually terete, rarely grass-like, perianth segments usually in 2 series of three each.
- pistils more than 1, distinct or separating at maturity into 3 to six follicles
- perianth inconspicuous, greenish, segments all alike, 3 to 6 carpels.
- flowers in a 3 to 12-flowered bracteate raceme, fruiting pedicels 12 to 25 mm long, follicles compressed, divergent, 1 or 2 seeded. - *Scheuchzeria palustris*
- flowers in more than 12-flowered, bracteate spikes, fruiting pedicels rarely up to 6 mm long, follicles neither compressed nor divergent, 1 seeded, no perianth, pistillate flowers axillary, perfect flowers in pedunculate spikes, long styles, fruit 3-lobed. - *Lilaea scilloides*
- perianth showy, sepals green, petals white to pinkish, 10 or more carpals, fruit an achene, stamens 6 (9) or

over 15, inflorescence a bracteate panicle or raceme. - Alismataceae

- leaves basally sagittate or hastate, 7 to 25 stamens, achenes densely packed on the receptacle. - *Sagittaria*
- mature achenes 2.0 to 2.5 mm long with a beak less than 0.5 mm long pointing forward from the tip of the achene, monoecious. - *Sagittaria cuneata*
- mature achenes (2.5) 3.0 to 3.5 (4.0) mm long with a beak about 1 mm long at right angles to the body of the achene, monoecious or dioecious. - *Sagittaria latifolia*
- leaves gradually tapered or cordate at the base; broadly ribbon-like if submersed.
- leaves completely submersed in deep water or floating on the surface, ribbon-like or tapered at the base and oval, flaccid, plants generally sterile. - *Sagittaria*
- leaves emergent and erect, ovate to lanceolate, stamens 6 (9), achenes in a single whorl on the receptacle. - *Alisma*
- leaf blades lanceolate to ovate, scapes longer than the leaves (petiole and blade), achenes centrally grooved apically. - *Alisma* plantago-aquatica
- leaf blades narrowly lanceolate to linear, scapes shorter than the leaves (petiole plus blade), achenes 2-grooved at the tip. - *Alisma gramineum*
- compound pistil, 1 or 3 celled, fruit not a follicle.
- plants grass like, flowers inconspicuous, perianth greenish, purplish or brownish, segments all alike. - *Juncus supiniformis*
- plants not grass-like, flowers showy, blue or purple, 6 stamens, long-petiolate, arrow-shaped, hastate leaves, perianth fused, inflorescence subtended by 2 spathes, introduction. - *Pontederia cordata*
- leaves not long and grass-like with parallel veins, flower parts not in 3's, usually in 5's, flowers yellow. - Dicotyledons (Ranunculaceae)
- leaves rounded to reniform with crenate margins, sepals but no petals, stem becoming creeping and nodally

rooting, flowers solitary on long stalks. - *Caltha palustris*

- leaves rhombic to chordate-rotund, all in a basal cluster, plants stoloniferous, petals with a basal pocket or nectary scale, flowers few to solitary on long stalks. - *Ranunculus cymbalaria*

Part 7: Marine Plants

- plants generally subtidal (may be briefly exposed at the lowest tides), long, narrow, grass-like leaves, many flowers embedded in a fleshy spadix, no elongate, coiled peduncle when fruiting. - Zosteraceae
- leaf sheaths deciduous, sometimes leaving a few scaly parts behind, leaf blades thin and translucent, rhizome with elongate internodes (1 to 3 cm long or more), 2 thin roots at each internode, monoecious, spadix border projections inconspicuous, usually established on sand or mud. - *Zostera*
- leaves 3-veined, 1 to 1.5 mm wide, sheaths split to the base, rare and probably introduced from Asia, known only from Boundary Bay and Tsawwassen. - *Zostera japonica*
- leaves 5-veined or more, 1.5 to 12 mm wide, sheaths on sterile shoots are closed at the base, widespread native species. - *Zostera marina*
- basal portions of the leaf sheaths decay with age to bundles of fine, woolly fibres, leaf blades leathery, rhizome has short, thick internodes with 2 or more thick roots at each internode, dioecious, spadix bordered by conspicuous flap-like projections, usually established rocks. - Phyllospadix
- fertile stems branched, 4 to 12 dm long, spathes usually paired at the nodes. - *Phyllospadix torreyi*
- fertile stems unbranched, 0.5 to 4 dm long, spathes usually solitary at the nodes.
- leaves with 3, rarely 5, veins, margins entire. - *Phyllospadix scouleri*
- leaves with 5 or 7 veins, margins toothed towards the apex. - *Phyllospadix serrulatus*

- plants submersed in brackish ditches and lagoons, long, narrow, grass-like leaves, two naked flowers on a short spadix arising from a sheath of leaf bases, 2 stamens and 4 pistils which become a cluster of fruits, peduncle and stipes elongate and coil at maturity, submersed. - *Ruppia maritima*

Classification of Aquatics by Life-form and Habitat

From an ecological point of view the growth-form of a plant and its usual habitat are often more important than the specific identification. Communities can often be distinguished by the growth-form of the plants present, which is constant world-wide, while the species may vary from place to place. Due to the morphological plasticity of aquatic plants no classification can be more than approximate and many exceptions will be found. A number of such classification schemes exist for various purposes. The following one may prove useful in reducing the number of choices when trying to identify an unknown plant, and in defining which group of plants normally inhabit certain zones. The key has been written to determine which group of plants is present, and under each group code mentioned in the key there is a list of species or genera which are applicable. Some plants will appear in more than one group. Generally only intact mature plants are keyed; fragments and seedlings would cause confusion.

Key to Mature Aquatic Plants of British Columbia: Based on Growth-forms and Habitat

- Floating freely in or on the water; attached only when stranded by lower water levels.
- Floating on the surface of the water.
- Large plants with distinct roots, stems and leaves; may have an obvious swollen part used as a floatation organ. - Group A
- Small thalloid plants with roots below; closely appressed to the surface of the water. - Group B
- Floating beneath the surface of the water.
- Found just under the surface tension of the water or free in the mid-water zone; often found tangled in other submersed or marginal vegetation. - Group C

- Found lying on or just above the sediment surface. - Group D
- Rooted or anchored in the sediment at all water levels. (fragments of *Elodea*, {*Egeria densa*}, *Najas flexilis*, *Ranunculus*, *Callitriche*, *Ceratophyllum*, {*Limnophila sessiliflora*} and *Myriophyllum* may be found adrift with Group C).
- All vegetative parts remain beneath the surface of the water. (several species begin the season submersed but as water levels drop end up in Group O; these are primarily plants of Group E and Group G but include a few members of Group I and Group J as well).
- Scapose plants with no leafy stem; leaves all arise from a basal rosette at the sediment surface.
- Leaves with a broad blade, or compound and resembling a clover leaf in *Marsilea vestita*. - Group E
- Leaves without a broad blade.
- Leaves very long and thin; limp. - Group F
- Leaves short, often fleshy or terete; stiff. - Group G
- Leafy stems with leaves scattered throughout the water column.
- Leaves all finely dissected, (in the Characeae there are no leaves, the analogous structures are branches and branchlets which are repeatedly forked). - Group H
- Leaves entire.
- Leaves filiform, long and narrow with no obvious blade. - Group I
- Leaves more or less broad with an obvious blade or wider portion. - Group J
- Vegetative parts may be floating or emergent from the water. (in the spring immature plants may not have reached the waters surface; late season flooding may inundate plants that were previously floating on the surface).
- Floating leaves present. (submersed leaves may also be present but generally no emergent leaves are present).
- Only floating leaves are present. - Group K

- Submersed and floating leaves are present.
- Leafy stems beneath the surface of the water with distinct floating leaves on the surface as well (the floating leaves of *Potamogeton alpinus* may not be distinctly different from the submersed leaves). - Group L
- Long, thin, limp, leaves arise beneath the surface of the water and the upper ends of these leaves float on the surface of the water. - Group M
- Leaves and/or stems emergent into the air. (generally no floating leaves are present except in *Sparganium* but some floating-leaf plants with stiff petioles may resemble Group O plants if water levels drop and the floating leaves are left emergent; *Nuphar* is an example).
- Leaves and stems lax and decumbent; sprawling over the surface of the water or adjacent shore. - Group N
- Leaves and stems rigid, erect and emergent. These are generally marginal plants of shallow water and may grow completely out of the water in late summer when water levels drop. - Group O

The Species Found in Each Growth-form Group

Group	*Genera and Species*	*Genera and Species*
A	{*Eichhornia crassipes*}	{*Limnobium laevigatum*}
	{*Limnobium spongia*}	{*Pistia stratiotes*}
	{*Salvinia*}	{*Trapa natans*}
B	*Azolla caroliniana*	*Azolla filiculoides*
	Azolla mexicana	*Lemna minor*
	Riccia fluitans	*Ricciocarpus natans*
	Spirodela polyrhizas	*Wolffia borealis*
	Wolffia columbiana	{*Wolffiella floridana*}
C	*Ceratophyllum demersum*	*Lemna trisulca*
	Utricularia gibba	*Utricularia minor*
	Utricularia vulgaris	*Wolffia borealis*
	Wolffia columbiana	{*Wolffiella floridana*}

(Cont...)

(Cont...)

Group	***Genera and Species***	***Genera and Species***
D	*Lemna trisulca*	*Utricularia gibba*
	Utricularia minor	*Utricularia vulgaris*
E	*Limosella aquatica*	*Marsilea vestita*
	Ranunculus flammula	*Ranunculus cymbalaria*
F	*Sagittaria*	*Vallisneria americana*
	Sparganium resemble this group but are not scapose; they have leafy stems	
G	*Isoetes*	*Lilaeopsis occidentalis*
	Lobelia dortmanna	{*Pilularia americana*}
	Ranunculus flammula	
H	{*Cabomba caroliniana*}	*Ceratophyllum*
	Chara	{*Limnophila sessiliflora*}
	Megalodonta beckii	*Myriophyllum*
	Nitella	*Ranunculus aquatilis*
	Ranunculus flabellaris	*Ranunculus sceleratus*
	Tolypella intricata	*Utricularia intermedia*
I	*Callitriche*	*Chara*
	Heteranthera dubia	*Hippuris vulgaris*
	Juncus supiniformis	*Najas flexilis*
	Nitella	*Phyllospadix* (marine)
	Potamogeton berchtoldii	*Potamogeton filiformis*
	Potamogeton foliosus	*Potamogeton friesii*
	Potamogeton obtusifolius	*Potamogeton pectinatus*
	Potamogeton pusillus	*Potamogeton robbinsii*
	Potamogeton strictifolius	*Potamogeton vaginatus*
	Potamogeton zosteriformis	*Ruppia maritima*
	Subularia aquatica	*Tolypella intricata*
	Zannichellia palustris	*Zostera* (marine)

J	*Crassula aquatica*	*Elatine rubella*
	{Egeria densa}	*Elodea*
	{Hydrilla verticillata}	*{Lagarosiphon major}*
	Ludwigia palustris	*Potamogeton crispus*
K	*Brasenia schreberi*	*Caltha natans*
	Nuphari	*Nymphaea*
	Nymphoides	*Polygonum amphibium*
	Potamogeton natans	*Ranunculus hyperboreus*
	Ranunculus lobbii	
L	*Callitriche*	*Potamogeton alpinus*
	Potamogeton amplifolius	*Potamogeton epihydrus*
	Potamogeton gramineus	*Potamogeton illinoensis*
	Potamogeton natans	*Potamogeton nodosus*
	Potamogeton oakesianus	*Potamogeton perfoliatus*
	Potamogeton praelongus	*Potamogeton richardsonii*
	Ranunculus aquatilis	*Ranunculus hyperboreus*
	Ranunculus lobbii	
M	*Glyceria*	*Sagittaria*
	Scirpus subterminalis	*Sparganium*
	Vallisneria americana	
N	*Callitriche*	*Heteranthera dubia*
	Juncus supiniformis	*Ludwigia palustris*
	Mimulus	*Myosotis*
	{Myriophyllum aquaticum}	*Nasturtium officinale*
	Veronica	
O	All the rooted, emergent plants belong to this group.	

Keys to the Species of the Growth-form Groups

Group A

- plants with long stems and leaves distributed along the stems. - *{Salvinia}*

- plants with all leaves in a compact rosette arising from a common point.
- petioles inflated to serve as floats.
- leaf bases cordate or abruptly narrowed to the petiole, flowers large, showy and purple, leaves glabrous and margins entire, rounded in outline, parallel veined. - {*Eichhornia crassipes*}
- leaf bases cuneate to truncate, flowers short-petiolate and inconspicuous, leaves rhombic and angular, apical margins at least toothed. - {*Trapa natans*}
- petioles not inflated.
- leaves petiolate, glabrous or nearly so, round to heart-shaped with a basal notch, spongy, green or marked with red, flat on the surface of the water. - {*Limnobium*}
- leaves have a thick spongy layer and are arched on the top surface, petiole short and leaves therefore closely clustered, leaves not marked. - {*Limnobium laevigatum*}
- leaves with a thin spongy layer, flat on the dorsal surface, petioles longer than the blades which are remote, leaves with dorsal red-brown markings. - {*Limnobium spongia*}
- leaves sessile or sub-sessile, pubescent, obovate and apically notched, leathery and grey-green, longitudinally ribbed., erect and emergent. - {*Pistia stratiotes*}

Group B

A compound microscope is required to positively identify species of *Azolla*.

- plants reduced to thalli with no stems or leaves, roots may be present.
- thalli not repeatedly and regularly forked, elongate to orbicular, flat or globose, flowering plants (duckweeds). - Lemnaceae
- one or more roots and nerves on each thallus (frond).
- each thallus bears two or more clustered roots from the base and 4 to 12 nerves. - *Spirodela polyrhiza*
- each thallus bears one root at the base and 1 to 3 nerves. - *Lemna*

- fronds oblong to lanceolate, 6 to 12 mm long and connected in small groups by stalks of the same length, matted, generally submersed, colonies. - *Lemna trisulca*
- fronds oval to round, less than 6 mm long, no stalks, solitary or in small attached groups, floating on the surface, usually 1 nerved, 2 to 4 mm long. - *Lemna minor*
- no roots or nerves on the thalli.
- thalli sub-globose to oval, up to 1 mm long. - *Wolffia*
- plants floating just below the surface, sub-globose, upper surface rounded, green, not puncticulate, 0.5 to 1.0 mm long. - *Wolffia columbiana*
- plants floating on the surface, ellipsoidal, upper surface flattish, white- or brown-punctilate, 0.5 to 1.2 mm long and about 1/2 as wide. - *Wolffia borealis*
- thalli long and narrow, sickle-shaped, several mm long but very narrow, flattened, submersed except at the base and aggregated into clusters. - {*Wolffiella floridana*}
- thalli repeatedly and regularly forked, bryophytes (liverworts).
- thalli broad, 2 to 3 times forked, tinged with purple, and fringed with tongue-shaped toothed scales. - *Ricciocarpus natans*
- thalli narrow and elongate, repeatedly dichotomously forked, not purple tinged nor fringed with scales. - *Riccia fluitans*
- stems with leaves and roots, leaves scale-like and overlapping, often reddish, ferns. - *Azolla*
- glochidia with several (usually 3 or more) septa, leaves at least 0.7 mm long, plant 1 to 3 cm in diameter, submersed leaf-lobes much larger than the upper lobes, glochidia unbranched. - *Azolla mexicana*
- glochidia with 1 (2) or no septa.
- glochidia without any septa, leaves papillose, oblong to ovate, about 1 mm long, megaspores coarsely roughened, plants 1 to 10 cm long, branches numerous and open, lower submersed leaf-lobe about as large as the upper lobe, glochidia unbranched. - *Azolla filiculoides*

- glochidia with 1 or 2 apical septa, leaves nearly smooth, sub-orbicular, about 0.5 mm long, megaspores finely roughened, plants usually less than 3 cm long, few branches and crowded, lower submersed leaf-lobe glabrous, larger and paler than the upper lobe, glochidia may be branched. - *Azolla caroliniana*

Group C

- plants with roots, stems and dissected leaves.
- plants with bladders on the alternate leaves. - *Utricularia*
- leaves divided into ultimately terete or threadlike segments.
- leaves divided into fewer than 5 final threadlike segments, leaf margins glabrous, bladders scarce on a small delicate plant usually floating at the surface or entangled in other rooted plants. - *Utricularia gibba*
- leaves 'pinnatifid', more than 20 terete final segments, hairy leaf margins, many bladders on the ordinary leaves, a robust plant to several meters long, usually lying on the sediment surface. - *Utricularia vulgaris*
- leaves divided di- or trichotomously into ultimately flattened segments.
- ordinary leaves generally with a few bladders, leaf margins glabrous, the terminal leaf segments are acuminate. - *Utricularia minor*
- ordinary leaves rarely, if ever, with bladders which are found on separate subterranean branches, leaf margins hairy, the ultimate leaf segment tips are awned. - *Utricularia intermedia*
- plants without bladders on the whorled leaves. - *Ceratophyllum*
- leaf segments sub-capillary, mostly entire, delicate and light green, in deeper water and not surfacing, achene with 3-5 lateral spines on each side, not a 'weedy' species. - *Ceratophyllum echinatum*
- leaf segments capillary to linear and flattened, serrate to coarsely toothed, plant usually coarse and robust,

dark green to almost black, usually surfacing, achene without lateral spines, 2 basal spines and 1 terminal spine only, a very 'weedy' species in eutrophic waters. - *Ceratophyllum demersum*

- plants reduced to thalli with no stems or leaves, roots may be present. - Lemnaceae
- one or more roots and nerves on each thallus (frond).
- each thallus bears two or more clustered roots from the base and 4 to 12 nerves. - *Spirodela polyrhiza*
- each thallus bears one root at the base and 1 to 3 nerves. - *Lemna*
- fronds oblong to lanceolate, 6 to 12 mm long and connected in small groups by stalks of the same length, matted, generally submersed, colonies. - *Lemna trisulca*
- fronds oval to round, less than 6 mm long, no stalks, solitary or in small attached groups, floating on the surface, usually 1 nerved, 2 to 4 mm long. - *Lemna minor*
- no roots or nerves on the thalli.
- thalli sub-globose to oval, up to 1 mm long. - *Wolffia*
- plants floating just below the surface, sub-globose, upper surface rounded, green, not puncticulate, 0.5 to 1.0 mm long. - *Wolffia columbiana*
- plants floating on the surface, ellipsoidal, upper surface flattish, white- or brown-punctilate, 0.5 to 1.2 mm long and about 1/2 as wide. - *Wolffia borealis*
- thalli long and narrow, sickle-shaped, several mm long but very narrow, flattened, submersed except at the base and aggregated into clusters. - {*Wolffiella floridana*}

Group D

- plants reduced to thalli with no stems or leaves, roots may be present. - Lemnaceae
- one or more roots and nerves on each thallus (frond).
- each thallus bears two or more clustered roots from the base and 4 to 12 nerves. - *Spirodela polyrhiza*
- each thallus bears one root at the base and 1 to 3 nerves. - *Lemna*

- fronds oblong to lanceolate, 6 to 12 mm long and connected in small groups by stalks of the same length, matted, generally submersed, colonies. - *Lemna trisulca*
- fronds oval to round, less than 6 mm long, no stalks, solitary or in small attached groups, floating on the surface, usually 1 nerved, 2 to 4 mm long. - *Lemna minor*
- no roots or nerves on the thalli.
- thalli sub-globose to oval, up to 1 mm long. - *Wolffia*
- plants floating just below the surface, sub-globose, upper surface rounded, green, not puncticulate, 0.5 to 1.0 mm long. - *Wolffia columbiana*
- plants floating on the surface, ellipsoidal, upper surface flattish, white- or brown-punctilate, 0.5 to 1.2 mm long and about 1/2 as wide. - *Wolffia borealis*
- thalli long and narrow, sickle-shaped, several mm long but very narrow, flattened, submersed except at the base and aggregated into clusters. - {*Wolffiella floridana*}
- plants with roots, stems and dissected leaves bearing bladders. - *Utricularia*
- leaves divided into ultimately terete or threadlike segments.
- leaves divided into fewer than 5 final threadlike segments, leaf margins glabrous, bladders scarce on a small delicate plant usually floating at the surface or entangled in other rooted plants. - *Utricularia gibba*
- leaves 'pinnatifid', more than 20 terete final segments, hairy leaf margins, many bladders on the ordinary leaves, a robust plant to several meters long, usually lying on the sediment surface. - *Utricularia vulgaris*
- leaves divided di- or trichotomously into ultimately flattened segments.
- ordinary leaves generally with a few bladders, leaf margins glabrous, the terminal leaf segments are acuminate. - *Utricularia minor*
- ordinary leaves rarely, if ever, with bladders which are found on separate subterranean branches, leaf margins

hairy, the ultimate leaf segment tips are awned. - *Utricularia intermedia*

Group E

- ferns with peltate, 4-lobed, 'clover-like' leaves coiled in the bud, long petioles. - *Marsilea vestita*
- flowering plants, leaves not 4-lobed and 'clover-like', leaves not coiled in the bud, short-petiolate or sessile.
- leaves broad, lobed or crenate, cordate, flowers yellow, fruit a head of achenes. - *Ranunculus cymbalaria*
- leaves entire and linear.
- flowers white, fruit a capsule. - *Limosella aquatica*
- flowers yellow, fruit a head of achenes. - *Ranunculus flammula*

Group F

Juvenile or sterile specimens may be difficult to distinguish.

- plants rhizomatous but not tuberous, flowers small and inconspicuous, on the end of a very long coiled pedicel which retracts below the water surface in fruit, flaccid leaves with many small longitudinal veinlets and cross-septa. - *Vallisneria americana*
- plants rhizomatous and tuberous, flowers white and showy in emergent racemes. - *Sagittaria*
- leaves basally sagittate or hastate, stamens 7 to 25, achenes densely packed on the receptacle, plants emergent in shallow water.
- monoecious, mature achenes 2.0 to 2.5 mm long with a beak less than 0.5 mm long pointing forward from the tip of the achene. - *Sagittaria cuneata*
- monoecious or dioecious, mature achenes (2.5) 3.0 to 3.5 (4.0) mm long with a beak about 1 mm long at right angles to the body of the achene. - *Sagittaria latifolia*
- leaves gradually tapered or cordate at the base; broadly ribbon-like, completely submersed in deep water or floating on the surface, ribbon-like or tapered at the base and oval, flaccid, plants generally sterile. - *Sagittaria*

In *Sagittaria* the usual non-achene key characters of bract length and shape and pedicel lengths have been found to be inconsistent and sufficiently variable even within one population as to be virtually useless as key characters. Many flowering, but non-fruiting, collections are difficult to identify.

Group G

A compound or dissecting microscope is required to positively identify species of *Isoetes*. Magnification over 10x is needed to examine the megaspore surface.

- ferns, leaves coiled in the bud, hairy sporocarps on the rhizome. - {*Pilularia americana*}
- not ferns, leaves not coiled in the bud, no hairy sporocarps.
- leaves not with 4 transversely septate, longitudinal gas chambers, flowering plants.
- leaves reduced to elongate, narrow, entire, hollow, transversely septate phyllodes, flowers in short umbels. - *Lilaeopsis occidentalis*
- leaves not as above, flowers not umbellate.
- plants solitary but usually colonial, a large cluster of white roots, white flowers in scapose racemes. - *Lobelia dortmanna*
- plants stoloniferous, no large conspicuous cluster of white roots, flowers on pedicels, solitary from near the ends of the stolons, flowers yellow, fruit a head of achenes. - *Ranunculus flammula*
- leaves sharply terete, sheathing bases with spores on the inside face of the base, each with 4 transversely septate, longitudinal gas chambers, stems short and tuberous. - *Isoetes*
- corms 3-lobed, plants primarily terrestrial, grow in grassy, ephemeral pools which are wet in winter and dry in summer. - *Isoetes nuttallii*
- corms 2-lobed, plants primarily of lake bottoms but sometimes exposed in summer due to water level fluctuations.
- megaspores spiny.

- megaspores vary greatly in size and are often aborted, spines blunt and dense. (this is probably a hybrid of *I. echinospora* and *I. maritima*). - *Isoetes truncata*
- megaspores uniform in size, spines sharp, not crowded.
- plants flaccid, megaspore spines elongate with pointed tips, uniform in size, microspores smooth. - *Isoetes echinospora*
- plants rigid, megaspore spines stubby and blunt, shorter and denser along the ridges, microspores spiny. - *Isoetes maritima*
- megaspores smooth or with ridges but not spiny.
- lowermost leaves in 2 ranks, megaspores over 0.5 mm in diameter, microspores rugose and over 36μm long. - *Isoetes occidentalis*
- lowermost leaves spirally arranged, megaspores less than 0.5 mm in diameter, microspore spiny and less than 30μm long.
- base of the leaves blackened, hyaline wing margins of the leaves extending 1 to 5 cm above the sporangium, sporangia brown spotted. - *Isoetes howellii*
- base of the leaves green, hyaline wing margins of the leaves not extending more than 1 cm above the sporangium, sporangia without colour. - *Isoetes bolanderi*

Group H

- flowering plants with roots, stems and leaves. - General Key: Aquatic Plants of British Columbia-Part 2.
- algae, no roots or leaves, only branches, produces clusters of reddish spores. - Characeae
- branchlets simple, not further branched or segmented, coronula 5-celled. - *Chara*
- stem not corticate, stipulodes in 1 tier, and alternating with the branchlets, branchlets tipped with a corona, no spine cells. - *Chara braunii*
- stems, and at least basal branchlet segment, corticate.

- spine cells in fascicles, axial cortex haplostichous, 1 corticate, stipulodes in 2 tiers, spiny looking, found in saline or brackish waters. - *Chara canescens*
- spine cells solitary or geminate, axial cortex diplostichous or triplostichous, 2 or 3 corticate.
- axial cortex diplostichous, stipulodes in 2 tiers, spine cells absent or small, rarely geminate, quite variable and widespread in hard-water lakes, smooth but marl encrusted. - *Chara vulgaris*
- axial cortex triplostichous, stipulodes usually in 2 tiers but may be in 1 tier, obscure or absent, spine cells usually absent or obscure but may be clustered, variable and widespread in hard-water and soft-water lakes, smooth appearance. - *Chara globularis*
- branchlets compound, re-branched and/or segmented.
- branchlets simple but segmented like a string of sausages, end cells reduced and acute, heads coarse and like a bird's nest. - *Tolypella intricata*
- branchlets compound, re-branched, may be segmented or apparently so. - *Nitella*
- the end segment of the branchlets is 1-celled.
- heteroclemous, 2 kinds of branchlets in a whorl, 1-forked alternating with 1-celled, bright green, compact, branchlets swollen. - *Nitella clavata*
- homeoclemous, all branchlets in a whorl the same, generally 1-forked.
- fertile heads small and densely compact, no mucous, generally no terminal dactyls on the sterile branchlets. - *Nitella acuminata*
- fertile heads absent or loose and not compact.
- dactyl apices not long acuminate, may be acute, blunt or apiculate, (a small variety may have 2-forked branchlets occasionally). - *Nitella flexilis*
- dactyl apices are long acuminate, gametangia on branchlets, generally not marl covered. - *Nitella acuminata*

- the end segment of the branchlets is 2-or-more-celled, the end cells are very much smaller than the penultimate cell, branchlets furcate, heads terminal, branchlets homeoclemous.
- robust plants, over 15 cm high, axes over 600 microns in diameter, dactyls may be small. - Nitella furcata
- smaller plants, seldom over 15 cm tall, axes 200 to 450 microns in diameter, dactyls uniform.
- lowest branchlet node fertile, dactyls 2 to 3-celled, heads may have mucous, oospore membrane granular or felt-like. - *Nitella gracilis*
- lowest branchlet node sterile, dactyls 2-celled, heads without mucous, oospore membrane reticulate. - *Nitella tenuissima*

Group I

Mature fruits are generally needed for positive identification of *Callitriche.*

- algae, produce clusters of reddish spores, no flowers, roots or leaves. - Characeae
- branchlets simple, not further branched or segmented, coronula 5-celled. - *Chara*
- stem not corticate, stipulodes in 1 tier, and alternating with the branchlets, branchlets tipped with a corona, no spine cells. - *Chara braunii*
- stems, and at least basal branchlet segment, corticate.
- spine cells in fascicles, axial cortex haplostichous, 1-corticate, stipulodes in 2 tiers, spiny looking, found in saline or brackish waters. - *Chara canescens*
- spine cells solitary or geminate, axial cortex diplostichous or triplostichous, 2- or 3-corticate.
- axial cortex diplostichous, stipulodes in 2 tiers, spine cells absent or small, rarely geminate, quite variable and widespread in hard-water lakes, smooth but marl encrusted. - *Chara vulgaris*
- axial cortex triplostichous, stipulodes usually in 2 tiers but may be in 1 tier, obscure or absent, spine cells usually absent or obscure but may be clustered, variable

and widespread in hard-water and soft-water lakes, smooth appearance. - *Chara globularis*

- branchlets compound, re-branched and/or segmented.
- branchlets simple but segmented like a string of sausages, end cells reduced and acute, heads coarse and like a bird's nest. - *Tolypella intricata*
- branchlets compound, re-branched, may be segmented or apparently so. - *Nitella*
- the end segment of the branchlets is 1-celled.
- heteroclemous, 2 kinds of branchlets in a whorl, 1-forked alternating with 1-celled, bright green, compact, branchlets swollen. - *Nitella clavata*
- homeoclemous, all branchlets in a whorl the same, generally 1-forked.
- fertile heads small and densely compact, no mucous, generally no terminal dactyls on the sterile branchlets. - *Nitella acuminata*
- fertile heads absent or loose and not compact.
- dactyl apices not long acuminate, may be acute, blunt or apiculate, (a small variety may have 2-forked branchlets occasionally). - *Nitella flexilis*
- dactyl apices are long acuminate, gametangia on branchlets, generally not marl covered. - *Nitella acuminata*
- the end segment of the branchlets is 2-or-more-celled, the end cells are very much smaller than the penultimate cell, branchlets furcate, heads terminal, branchlets homeoclemous.
- robust plants, over 15 cm high, axes over 600 microns in diameter, dactyls may be small. - *Nitella furcata*
- smaller plants, seldom over 15 cm tall, axes 200 to 450 microns in diameter, dactyls uniform.
- lowest branchlet node fertile, dactyls 2- to 3-celled, heads may have mucous, oospore membrane granular or felt-like. - *Nitella gracilis*
- lowest branchlet node sterile, dactyls 2-celled, heads without mucous, oospore membrane reticulate. - *Nitella tenuissima*

- flowering plants with roots, stems and leaves.
- marine plants. - Zosteraceae
- leaf sheaths deciduous, sometimes leaving a few scaly parts behind, leaf blades thin and translucent, rhizome with elongate internodes (1 to 3 cm long or more), with 2 thin roots at each internode, monoecious, spadix border projections inconspicuous, if present, usually established on sand or mud. - *Zostera*
- leaves 3-veined, 1 to 1.5 mm wide, sheaths split to the base, rare and probably introduced from Asia, known from Boundary Bay and Tsawwassen. - *Zostera japonica*
- leaves 5-veined or more, 1.5 to 12 mm wide, sheaths on sterile shoots are closed at the base, widespread native species. - *Zostera marina*
- basal portions of the leaf sheaths decay with age to bundles of fine, woolly fibres, leaf blades leathery, rhizome has short, thick internodes with 2 or more thick roots at each internode, dioecious, spadix bordered by conspicuous flap-like projections, usually established only on hard or rocky substrates. - *Phyllospadix*
- fertile stems branched, 4 to 12 dm long, spathes usually paired at the nodes. - *Phyllospadix torreyi*
- fertile stems unbranched, 0.5 to 4 dm long, spathes usually solitary at the nodes.
- leaves with 3 (rarely 5) veins, margins entire. - *Phyllospadix scouleri*
- leaves with 5 or 7 veins, margins toothed towards the apex. - *Phyllospadix serrulatus*
- fresh water or brackish water plants.
- inflorescence an umbel on a long coiled pedicel, brackish water. - *Ruppia maritima*
- inflorescence not umbellate, pedicel not coiled in fruit.
- leaves in a basal cluster, flowers racemose. - *Subularia aquatica*
- leaves cauline.
- leaves whorled or opposite.

- leaves in whorls of 6 or more, submersed leaves to 5 cm long or more, flowers axillary, stem apex often emergent. - *Hippuris vulgaris*
- leaves opposite, submersed leaves 2 (2.5) cm long, flowers axillary, stem apices may form a rosette floating on the surface. - *Callitriche*
- fruit encircled by a conspicuous wing-like margin, leaf bases joined by winged ridges. - *Callitriche stagnalis*
- fruit not winged or with only a narrow wing at the tip, leaf bases various.
- leaves all linear, 1-nerved, light green, leaf bases not joined by a wing floral bracts absent, common species. - *Callitriche hermaphroditica*
- leaves various, upper often ovate and 3-nerved, bases joined by a wing-like ridge, floral bracts present.
- carpel face markings in regular vertical lines, fruit slightly wing-margined at the top and longer than broad. - *Callitriche verna*
- carpel face markings scattered, fruit not winged and as long as broad.
- fruits widest above the middle (obovate) leaves bidentate, midvein barely thickened at the end, emergent leaves may be over 5 mm wide, stems long. - *Callitriche heterophylla*
- fruits round or oblong, midvein thickened at the tip and protruding, leaves never over 5 mm wide, plants short and slender. - *Callitriche anceps*
- leaves alternate.
- plants stoloniferous and proliferous, forming a dense, often reddish, tangled mat on the surface, leaves auriculate, perianth brown. - *Juncus supiniformis*
- plants not stoloniferous, proliferous or mat-forming, leaves not auriculate.
- fruits a cluster of banana-shaped achenes in the leaf axils, female flowers in a cup-shaped sheath. - *Zannichellia palustris*

- fruits not a cluster of banana-shaped achenes, flowers not in cup-shaped sheaths.
- plants sprawling and rooted at the lower nodes at least, flower pedicels arising from the stipulate leaf axils, solitary yellow flowers from spathes, fruit a capsule. - *Heteranthera dubia*
- plants erect, not rooting at the nodes, flowers small, numerous and inconspicuous in terminal, pedunculate spikes, no spathes, fruit an achene. - *Potamogeton*

Group J

- leaves alternate.
- leaves sessile and clasping, crispate, finely serrulate, leaves oblong 3 to 8 cm long and 3 to 10 mm wide, flowers in a terminal emergent inflorescence. - *Potamogeton crispus*
- leaves not clasping or crispate, inflorescence axillary.
- leaves strongly recurved, serrate or toothed, sessile, linear to linear-lanceolate. - {*Lagarosiphon major*}
- leaves straight, entire, petiolate, ovate-elliptic. - *Ludwigia palustris*
- leaves opposite or whorled.
- leaves opposite, entire, often fleshy, flowers solitary in the leaf axils.
- flowers 4-merous, 4 erect purplish follicles with 6 to 12 seeds. - *Crassula aquatica*
- flowers 2- or 3-merous, capsules with extensively pitted, curved seeds. - *Elatine rubella*
- leaves whorled, may be toothed, membranous. - *Hydrocharitaceae*
- leaves very long and narrow in basal rosettes, pedicel arises from the rosette and reaches the surface, spirally coiled. - *Vallisneria americanus*
- leaves short, up to 4 cm long, arising from the elongated stem.
- leaf arrangement irregular, margins serrate or toothed, tip with 2 enlarged spines - {*Lagarosiphon major*}

- leaves arranged in regular whorls.
- leaf margins serrate or toothed.
- leaf tip with 2 enlarged spines, no turions, 3 or more strongly recurved leaves up to 3 cm long per whorl. - {*Lagarosiphon major*}
- no enlarged spines on the leaf tip, turions present on the rhizome or on the stem, 3 to 8 (12) nearly straight leaves up to 2.5 cm long per whorl. - {*Hydrilla verticillata*}
- leaf margins entire.
- basal leaves in whorls of 3 but upper ones up to 6 per whorl and to 4 cm long, 2 to 3 flowers in the staminate spathes, petals to 10 mm. - {*Egeria densa*}
- basal leaves in pairs, the upper ones usually in whorls of 3 and up to 2.5 cm long, flowers solitary in the staminate spathes, petals up to 5 mm long. - *Elodea*
- leaves in whorls of 3.
- leaves rarely over 1.5 cm long.
- leaves 2 (1 to 4) mm wide, tapered abruptly to a blunt point, staminate flowers stalked and persistent. - *Elodea canadensis*
- leaves 1.5 (0.3 to 1.5) mm wide, tapered to a slender point, staminate flowers sessile and deciduous at anthesis. - *Elodea nuttallii*
- leaves (1.7) 2.0 to 2.6 cm long. - {*Elodea longivaginata*}
- leaves in the upper and middle part of the stem in pairs, lower ones irregular or alternate, (1.7) 2.0 to 2.6 cm long. - {*Elodea longivaginata*}

Group K

- leaves peltate, submersed portions of the plants mucilaginous. - *Brasenia schreberi*
- leaves not peltate, petiole attached at the leaf margin, plants not mucilaginous.
- leaves entire, over 1 cm long, flowers not yellow.
- leaves orbicular, bases hastate or cordate.

- fruit is a berry, 4 to 6 green sepals, many white, yellow or red petals and many stamens, rhizome massive. - Nymphaeaceae
- fruits a cluster of follicles, 5 or 6 small white sepals, no petals and up to 20 stamens, slender creeping stolons. - *Caltha natans*
- leaves not orbicular, longer than broad, bases cordate, reniform or tapered.
- leaves cordate or reniform at the base, ovate to elliptic.
- leaves acute at the apex and cordate at the base, usually reddish, petioles reflexed at the leaf junction, leaf margin entire. - *Potamogeton natans*
- leaves rounded at the apex and reniform to cordate at the base, greenish, leaf margin toothed. - *Caltha natans*
- leaves obtuse to basally tapered, lanceolate to narrowly-elliptic, apex acute, green, flowers rose or pink in compact panicles. - *Polygonum amphibium*
- leaves lobed, up to about 1 cm long, flowers yellow, fruit a head of achenes. - *Ranunculus hyperboreus*

Group L

Mature fruits are generally needed for positive identification of *Callitriche*.

- leaves opposite, entire, 1 or 3 nerved, up to 2 (2.5) cm long, no stipules, flowers sessile in the axils and submersed. - *Callitriche*
- fruit encircled by a conspicuous wing like margin, leaf bases joined by winged ridges. - *Callitriche stagnalis*
- fruit not winged or with only a narrow wing at the tip, leaf bases various.
- leaves all linear, 1-nerved, light green, leaf bases not joined by a wing, floral bracts absent, common species. - *Callitriche hermaphroditica*
- leaves various, upper often ovate and 3-nerved, bases joined by a wing-like ridge, floral bracts present.
- carpel face markings in regular vertical lines, fruit slightly wing-margined at the top and longer than broad. - *Callitriche verna*

- carpel face markings scattered, fruit not winged and as long as broad.
- fruits widest above the middle (obovate) leaves bidentate, midvein barely thickened at the end, emergent leaves may be over 5 mm wide, stems long. - *Callitriche heterophylla*
- fruits round or oblong, midvein thickened at the tip and protruding, leaves never over 5 mm wide, plants short and slender. - *Callitriche anceps*
- leaves alternate, entire or lobed, usually more than 3 veins, stipulate or with an inflated petiole base, flowers pedicellate, emergent.
- floating leaves entire, flowers inconspicuous in terminal, emergent inflorescences, leaves generally well over 2 cm long, stipules present. - *Potamogeton*
- floating leaves lobed, flowers solitary, axillary, yellow or white, leaves from 1 to 2 cm long, base of petiole inflated and sheathing. - *Ranunculus*
- flowers white, submersed leaves dissected into ultimately filiform segments.
- two to 7 glabrous, beakless achenes, plants and receptacle glabrous, pedicels in the axils of ternately lobed floating leaves, leaves 2 to 3 times divided into 8 to 12 segments. - *Ranunculus lobbii*
- 10 to 80 glabrous or hirsute achenes, plants and receptacles hirsute.
- 30 to 80 achenes, flowers may have yellow bases, leaf blades sessile on the stipular base. - *Ranunculus circinatus*
- ten to 25 achenes, flowers all white, leaf blades petiolate. - *Ranunculus aquatilis*
- flowers yellow, submersed leaves, if present, simple to ternately divided.
- leaves simple and entire to serrate, may be shallowly lobed, achenes glabrous.
- achenes 50 to 200 in a columnar head, longitudinally ribbed. - *Ranunculus cymbalaria*

- achenes 5 to 60 in a globular head, not longitudinally ribbed, stems decumbent to prostrate, rooting at the nodes, nectary scales usually broader than long. - *Ranunculus flammula*
- leaves all, or in part, lobed, parted or ternately dissected to filiform segments, achenes pubescent.
- annuals, erect, not nodally rooting, achene beaks inconspicuous. - *Ranunculus sceleratus*
- perennials, floating or reclining, nodally rooting, achene beaks conspicuous.
- leaves deeply lobed or parted, achenes not corky margined.
- leaves deeply 3 parted with narrow lobes distally acute, achene beaks 1/4 the length of the achenes, receptacles hirsute, 5 sepals and petals. - *Ranunculus gmelinii*
- leaves 3-lobed, distally rounded, achene beaks 1/10 the length of the achenes, receptacles glabrous, 3 sepals and petals. - Ranunculus hyperboreus
- leaves, at least the submersed ones, 3 to 5 times ternately dissected into filiform segments less than 2 mm wide, achenes corky margined. - *Ranunculus flabellaris*

Group M

- leaves filiform or terete, usually under 3 mm wide, a solitary sub-terminal spikelet in the axil of a prominent bract. - *Scirpus subterminalis*
- leaves usually broader than 3 mm, not filiform or terete.
- swollen nodes, closed sheathing leaf bases, prominent ligules, hollow culms, articulated panicles, leaves often wavy-margined on the surface of the water, grasses. - *Glyceria*
- nodes not swollen, no closed sheathing leaf base or ligules, stems not hollow, inflorescence not an articulated panicle.
- leaves thin and flat, 3 to 10 mm wide, with numerous longitudinal veinlets and cross septa, pistillate flowers solitary on the end of a long pedicel which coils and

retracts beneath the surface in fruit. - *Vallisneria americana*

- flowers in emergent inflorescences.
- individual flowers numerous and inconspicuous in heads, fruit of achenes. - *Sparganium*
- individual flowers few with showy petals in whorled bracteate racemes. - *Sagittaria*
- leaves basally sagittate or hastate, stamens 7 to 25, achenes densely packed on the receptacle, plants emergent in shallow water.
- monoecious, mature achenes 2.0 to 2.5 mm long with a beak less than 0.5 mm long pointing forward from the tip of the achene. - *Sagittaria cuneata*
- monoecious or dioecious, mature achenes (2.5) 3.0 to 3.5 (4.0) mm long with a beak about 1 mm long at right angles to the body of the achene. - *Sagittaria latifolia*
- leaves gradually tapered or cordate at the base; broadly ribbon-like, submersed in deep water or floating on the surface, ribbon-like or tapered at the base and oval, flaccid, plants generally sterile. - *Sagittaria*

Group N

Mature fruits are generally needed for positive identification of *Callitriche*.

- leaves finely dissected, pale yellowish-green, flowers small, in axillary whorls. - {*Myriophyllum aquaticum*}
- leaves not dissected.
- leaves, at least the upper ones, compound, apical lobe much larger than the laterals, flowers white in racemes, fruit an elongate silique. - *Nasturtium officinale*
- leaves simple, fruits not siliques.
- leaves opposite.
- leaves entire, flowers sessile in the axils.
- leaves elliptic to ovate, 2 to 6 cm long, petiolate, flowers 4 merous and sepaloid, fruit a capsule. - *Ludwigia palustris*

- leaves linear or spathulate to ovate, up to 2 (2.5) cm long, flowers naked, 1 stamen, 2 styles, 2-lobed achene-like fruit, often winged. - *Callitriche*
- fruit encircled by a conspicuous wing-like margin, leaf bases joined by winged ridges. - *Callitriche stagnalis*
- fruit not winged or with only a narrow wing at the tip, leaf bases various.
- leaves all linear, 1-nerved, light green, leaf bases not joined by a wing, floral bracts absent, common species. - *Callitriche hermaphroditica*
- leaves various, upper often ovate and 3-nerved, bases joined by a wing-like ridge, floral bracts present.
- carpel face markings in regular vertical lines, fruit slightly wing-margined at the top and longer than broad. - *Callitriche verna*
- carpel face markings scattered, fruit not winged and as long as broad.
- fruits widest above the middle (obovate) leaves bidentate, midvein barely thickened at the end, emergent leaves may be over 5 mm wide, stems long. - *Callitriche heterophylla*
- fruits round or oblong, midvein thickened at the tip and protruding, leaves never over 5 mm wide, plants short and slender. - *Callitriche anceps*
- leaves dentate.
- leaves broadly ovate to obovate, coarsely toothed, flowers solitary on long pedicels in the leaf axils, flowers pink-purple or yellow, large, showy and zygomorphic. - *Mimulus*
- corollas pink-purple, stems over 3 dm tall, leaves 3 to 10 cm long. - *Mimulus lewisii*
- corollas yellow with reddish or purplish markings, perennials with the upper calyx teeth larger than the others.
- rhizomatous and often stoloniferous, flowers usually 1 to 5, corollas 2 to 4 cm long. - *Mimulus tilingii*

- stoloniferous, rarely rhizomatous, usually 5 or more flowers, corollas usually less than 2 cm long. - *Mimulus guttatus*
- leaves linear-lanceolate to ovate, finely toothed, many pedicellate flowers in axillary racemes, flowers blue, small and zygomorphic. - *Veronica*
- leaves all short-petiolate. - *Veronica beccabunga*
- leaves, at least those on the middle and upper portions of the flowering shoots, sessile.
- capsule flattened, notched, wider than high, 5 to 9 seeds per locule, 1 to 2 mm long, leaves 4 to 20 times as long as wide. - *Veronica scutellata*
- capsule turgid, scarcely notched, barely wider than high, many seeds less than 0.5 mm long, leaves 1.5 to 5 times as long as wide.
- leaves 1.5 to 3 times as long as wide, fruiting pedicels ascending, capsules higher than wide if not equal, flowers blue or violet. - *Veronica anagallis-aquatica*
- leaves 2.5 to 5 times as long as wide, fruiting pedicels spreading, capsules wider than high, flowers white to pink or pale blue. - *Veronica catenata*
- leaves alternate.
- leaves filiform, terete, auriculate, plants proliferous and often reddish, 3 or 6 stamens, flowers brown, the perianth is undifferentiated. - *Juncus supiniformis*
- leaves linear to lanceolate, not filiform, terete or auriculate, plants not proliferous or reddish, stamens 3 or 5, perianth differentiated and showy.
- linear stipulate leaves, flowers yellow, stamens 3, solitary, long-pedicellate flowers from a spathe. - *Heteranthera dubia*
- corolla limb 5 to 10 mm wide, style as long as or longer than the nutlets, stoloniferous, mostly erect and not creeping or recumbent. - *Myosotis scorpioides*

Group O

These plants are keyed out in Parts 4, 5, and 6 of the General Key.

Chapter 7

Keys to Aquatic Genera and Species within Families

Key to the Alismataceae of British Columbiae

In *Sagittaria* the usual non-achene key characters of bract length and shape and pedicel lengths have been found to be inconsistent and sufficiently variable even within one population as to be virtually useless as key characters. Many flowering, but non-fruiting, collections are difficult to identify.

- leaves basally sagittate or hastate, stamens 7-25, achenes densely packed on the receptacle. - *Sagittaria*
- monoecious, mature achenes 2.0-2.5 mm long with a beak less than 0.5 mm long pointing forward from the tip of the achene. - *Sagittaria cuneata*
- monoecious or dioecious, mature achenes (2.5) 3.0-3.5 (4.0) mm long with a beak about 1 mm long at right angles to the body of the achene. - *Sagittaria latifolia*
- leaves gradually tapered or cordate at the base; broadly ribbon-like if submersed.
- leaves completely submersed in deep water or floating on the surface, ribbon-like or tapered at the base and oval, flaccid, plants generally sterile. - *Sagittaria*
- leaves emergent and erect, ovate to lanceolate, stamens 6 (9), achenes in a single whorl on the receptacle. - *Alisma*
- leaf blades lanceolate to ovate, scapes longer than the leaves (petiole plus blade), achenes centrally grooved at the top. - *Alisma plantago-aquatica*

- leaf blades narrowly lanceolate to linear, scapes shorter than the leaves (petiole plus blade), achenes 2-grooved at the tip. - *Alisma gramineum*

Key to the Aquatic Apiaceae of British Columbia

- leaves with compound blades, plants erect and rooting at lower nodes only, umbels compound, base of the stem without transverse septa, roots not tuberous-thickened, ribs of the fruit prominent, corky-thickened, leaves with defined leaflets not dissected into narrow segments, primary lateral veins of the leaflets not directed in any particular direction, plants erect, calyx teeth tiny or absent. - *Sium sauve*
- leaves with simple blades or bladeless, plants creeping and rooting at all the nodes, umbels simple.
- leaves bladeless, reduced to linear or spathulate, septate stalks, leaves and umbels clustered at the nodes. - *Lilaeopsis occidentalis*
- leaves with blades, solitary on long petioles from the nodes, umbels solitary on long peduncles from the same nodes. - *Hydrocotyle*
- leaves sub-orbicular to reniform, crenate or shallowly lobed, centrally peltate. - *Hydrocotyle verticillata*
- leaves 5 to 6 lobed nearly or quite to the middle, not peltate. - *Hydrocotyle ranunculoides*

The lower-most leaves of many marginal Umbellifers may be under water, especially in the spring and may then be finely dissected with filiform segments. *Sium sauve* is especially prone to being found completely under water in the spring with all leaves filiform-dissected. Such plants will not key out properly in the general key. *Cicuta* is poisonous, use care when cutting open tubers to check for transverse septa. Wash your hands and your knife afterwards.

Key to the Aquatic Araceae of British Columbia

- plants free-floating, leaves thick and leathery, hairy, sub-sessile and cuneate, in a rosette, tropical aquarium and garden pool plants, not yet known to overwinter outdoors in BC. - {*Pistia stratiotes*}

- plants bottom-rooted, leaves glabrous, petiolate or linear and reed-like, native, leaves bifacial with broad blades and petioles, spadix basal and stalked, distinct showy spathes, leaves ovate and up to 30 cm long, bases cordate, spathe white and up to 6 cm long, spadix short, clavate with large red berries. - *Calla palustris*

Key to the Cabombaceae of British Columbia

- leaves peltate and entire, floating on the surface, plant mucilaginous, flowers red, native. - *Brasenia schreberi*
- leaves dissected and submersed, plants not mucilaginous, flowers with white sepals and yellow petals, aquarium plant, not yet found out of cultivation. - {*Cabomba caroliniana*}

Key to the Characeae of British Columbia

- branchlets simple, not further branched or segmented, coronula 5-celled. - *Chara*
- stem not corticate, stipulodes in 1 tier, and alternating with the branchlets, branchlets tipped with a corona, no spine cells. - *Chara braunii*
- stems, and at least basal branchlet segment, corticate.
- spine cells in fascicles, axial cortex haplostichous, 1-corticate, stipulodes in 2 tiers, spiny looking, found in saline or brackish waters. - *Chara canescens*
- spine cells solitary or geminate, axial cortex diplostichous or triplostichous, 2 or 3 corticate.
- axial cortex diplostichous, stipulodes in 2 tiers, spine cells absent or small, rarely geminate, quite variable and widespread in hard-water lakes, smooth but marl encrusted. - *Chara vulgaris*
- axial cortex triplostichous, stipulodes usually in 2 tiers but may be in 1 tier, obscure or absent, spine cells usually absent or obscure but may be clustered, variable and widespread in hard-water and soft water lakes, smooth appearance. - *Chara globularis*
- branchlets simple but segmented like a string of sausages, end cells reduced and acute, heads coarse and like a bird's nest. - *Tolypella intricata*

- branchlets compound, re-branched, may be segmented or apparently so. - *Nitella*
- the end segment of the branchlets is 1-celled.
- heteroclemous, 2 kinds of branchlets in a whorl, 1-forked alternating with 1-celled, bright green, compact, branchlets swollen. - *Nitella clavata*
- homeoclemous, all branchlets in a whorl the same, generally 1-forked.
- fertile heads small and densely compact, no mucous, generally no terminal dactyls on the sterile branchlets. - *Nitella acuminata*
- fertile heads absent or loose and not compact.
- dactyl apices not long acuminate, may be acute, blunt or apiculate, (a small variety may have 2-forked branchlets occasionally). - *Nitella flexilis*
- dactyl apices are long acuminate, gametangia on branchlets, generally not marl covered. - *Nitella acuminata*
- the end segment of the branchlets is 2-or-more-celled, the end cells are very much smaller than the penultimate cell, branchlets furcate, heads terminal, branchlets homeoclemous.
- robust plants, over 15 cm high, axes over 600 microns in diameter, dactyls may be small. - *Nitella furcata*
- smaller plants, seldom over 15 cm tall, axes 200 to 450 microns in diameter, dactyls uniform.
- lowest branchlet node fertile, dactyls - to 3-celled, heads may have mucous, oospore membrane granular or felt-like. - *Nitella gracilis*
- lowest branchlet node sterile, dactyls 2-celled, heads without mucous, oospore membrane reticulate. - *Nitella tenuissima*

Key to the Aquatic and Wetland Cyperaceae of British Columbia

- achenes enclosed in a perigynium, flowers unisexual.
- perigynium open, margins free (not covered in this key). - *Kobresia*

- perigynium closed, a dorsal suture may be evident (not covered in this key). - *Carex*
- achenes not enclosed in a perigynium, flowers perfect.
- scales distichous, in 2 vertical ranks, perianth of 6-9 bristles, stem leafy, upper leaves with axillary inflorescences. - *Dulichium arundinaceum*
- scales of the spikelet spirally arranged.
- styles basally thickened and persistent on the achenes as a tubercle.
- several to many spikelets, 1 or 2 fertile flowers or achenes per spikelet, 10 to 12 perianth bristles (not covered in this key). - *Rhynchospora*
- spikelet solitary, several to many fertile flowers or achenes per spikelet, up to 6 perianth bristles (not covered in this key). - *Eleocharis*
- styles not thickened, achenes may be apiculate.
- achenes subtended by more than 10 conspicuous and elongate white bristles, stigmas trifid and achenes trigonal (not covered in this key). - *Eriophorum*
- achenes subtended by 8 or fewer inconspicuous perianth bristles, stigmas bifid or trifid, achenes lenticular or trigonal.
- Spikelets with slightly modified lower scales, not leafy.
- Lower scales of the spikelets short, not more than half as long as the spikelet (not covered in this key). - *Eleocharis*
- Lower scales of the spikelets long, nearly as long or longer than the spikelets (not covered in this key. - *Trichophorum*
- Spikelets with subtending leafy bracts. (bristles present and/or achenes apiculate, perennials, involucral bract solitary, not leafy or spreading, apparently an extension of the culm.) - *Scirpus*
- spikelets numerous in branching inflorescences, often sessile in clusters at the ends of branches, species complex. - *Scirpus lacustris*

- spikelets solitary or few, sessile in a sessile cluster, spikelets solitary or rarely 2, stems flaccid and the apices floating on the surface of the water, leaves and stems filiform. - *Scirpus subterminalis*

Key to the Hydrocharitaceae of British Columbia

- leaves very long and narrow in basal rosettes, pedicel arises from the rosette and reaches the surface, spirally coiled - *Vallisneria americanus*
- leaves short, up to 4 cm long, arising from the elongated stem.
- leaf arrangement irregular, margins serrate or toothed, tip with 2 enlarged spines - {*Lagarosiphon major*}
- leaves arranged in regular whorls.
- leaf margins serrate or toothed.
- leaf tip with 2 enlarged spines, no turions, 3 or more strongly recurved leaves up to 3 cm long per whorl. - {*Lagarosiphon major*}
- no enlarged spines on the leaf tip, turions present on the rhizome or on the stem, 3 to 8 (12) nearly straight leaves up to 2.5 cm long per whorl. - {*Hydrilla verticillata*}
- leaf margins entire.
- basal leaves in whorls of 3 but upper ones up to 6 per whorl and up to 4 cm long, 2 to 3 flowers in the staminate spathes, petals to 10 mm.{*Egeria densa*}
- basal leaves in pairs, the upper ones usually in whorls of 3 and up to 2.5 cm long, flowers solitary in the staminate spathes, petals up to 5 mm long. - *Elodea*
- leaves in whorls of 3.
- leaves rarely over 1.5 cm long.
- leaves 2 (1 to 4) mm wide, tapered abruptly to a blunt point, staminate flowers stalked and persistent. - *Elodea canadensis*
- leaves 1.5 (0.3 to 1.5) mm wide, tapered to a slender point, staminate flowers sessile and deciduous at anthesis. - *Elodea nuttallii*

- leaves (1.7) 2.0 to 2.6 cm long. - {*Elodea longivaginata*}
- leaves in the upper and middle part of the stem in pairs, lower ones irregular or alternate, (1.7) 2.0 to 2.6 cm long. - {*Elodea longivaginata*}

Key to the Lemnaceae of British Columbia

- one or more roots and nerves on each thallus (frond).
- each thallus bears two or more clustered roots from the base and 4 to 12 nerves. - *Spirodela polyrhiza*
- each thallus bears one root at the base and 1 to 3 nerves. - *Lemna*
- fronds oblong to lanceolate, 6 to 12 mm long and connected in small groups by stalks of the same length, matted, generally submersed, colonies. - *Lemna trisulca*
- fronds oval to round, less than 6 mm long, no stalks, solitary or in small attached groups, floating on the surface, usually 1 nerved, 2 to 4 mm long. - *Lemna minor*
- no roots or nerves on the thalli.
- thalli sub-globose to oval, up to 1 mm long. - *Wolffia*
- plants floating just below the surface, sub-globose, upper surface rounded, green, not puncticulate, 0.5 to 1.0 mm long. - *Wolffia columbiana*
- plants floating on the surface, ellipsoidal, upper surface flattish, white- or brown-punctilate, 0.5 to 1.2 mm long and about 1/2 as wide. - *Wolffia borealis*
- thalli long and narrow, sickle-shaped, several mm long but very narrow, flattened, submersed except at the base and aggregated into clusters. - {*Wolffiella floridana*}

Key to the Menyanthaceae of British Columbia

- leaves trifoliate and usually entire, corolla with a mass of fimbriate scales, spicate. - *Menyanthes trifoliata*
- leaves simple, hastate or cordate, the white or yellow flowers are either solitary or clustered and appearing to arise from the petiole or base of the leaf, leaves small to about 15 cm long. - *Nymphoides*

- flowers yellow, solitary and axillary. - {*Nymphoides peltata*}
- flowers white, clustered at the base of the leaf.
- leaf blades ovate and mostly under 5 cm long, not dark-punctate beneath, seeds smooth. {*Nymphoides cordatum*}
- leaf blades orbicular and 8 to 15 cm in diameter, dark-punctate or pitted beneath, seeds glandular, warty. - {*Nymphoides aquatica*}

Key to the Nymphaeaceae of British Columbia

- large discoid stigma, superior ovary, about 6 large yellowish conspicuous sepals, flowers yellow, leaves flat, dark green, entire margined and leathery, native, large robust plants. - *Nuphar*
- stamens reddish, petiole round in cross section, primarily western and coastal BC. - *Nuphar polysepalum*
- stamens yellow, petiole flattened in cross section, primarily eastern BC. - *Nuphar variegatum*
- several to many spreading stigmas, partially inferior ovary, 4 greenish inconspicuous sepals, flowers red, pink, white and yellow, leaves often suffused with red especially below and often wavy margined but not leathery, many horticultural varieties are naturalized and cause confusion. - *Nymphaea*
- petals yellow or red, pink or reddish.
- about 25 yellow petals, flowers 6 to 10 cm in diameter. - *Nymphaea mexicana*
- petals red, pink, reddish (horticultural varieties). - *Nymphaea*
- petals white.
- flowers 7 to 12 cm in diameter, over 20 petals, over 8 styles.
- white petals gradually transitional to the staminodia and stamens, spirally arranged, stigmatic disc with 8 to 24 radial stigmas, not fragrant. - *Nymphaea alba*
- 20 to 30 thick white elliptic petals followed by 70 to 100 yellow stamens and staminodia, about 20 styles, flowers

very fragrant, open in the early morning and close later in the day. - *Nymphaea odorata*

- carpellary appendages 3 mm long or more, flaccid and purplish. - *Nymphaea tetragona*
- carpellary appendages 1.5 mm long or less, stiff and greenish. - *Nymphaea leibergii*

Key to the Pontederiaceae of British Columbia

- plants erect and emergent, fruit a 1-seeded nut enclosed by the base of the perianth tube, locally introduced and usually in cultivation, 6 stamens, inflorescence geniculate at anthesis, flowers blue. - {*Pontederia cordata*}
- plants submersed, sprawling or floating on the surface, fruit a 3-locular capsule.
- floating on the surface, petiole inflated, conspicuous purple flowers, 6 stamens, introduced in garden pools but not surviving outdoors over winter. - {*Eichhornia crassipes*}
- submersed or sprawling at the surface, petiole not inflated, inconspicuous yellow flowers, 3 stamens, native. - *Heteranthera dubia*

Key to the Aquatic Ranunculaceae of British Columbia

- leaves dissected, compound or lobed, petals white or yellow, sepals green and often deciduous, fruit a head of achenes. - *Ranunculus*
- leaves simple, petals absent, sepals showy and white or yellow, fruit a cluster of elongate follicles, leaves finely crenate margined. - *Caltha*
- stems stout, erect or sprawling, leaves erect, over 5 cm wide, flowers yellow. - *Caltha palustris*
- stems slender, creeping or floating, leaves floating, under 5 cm wide, flowers white. - *Caltha natans*

Key to the Aquatic Scrophulariaceae of British Columbia

- petal tube very short and the lobes apparently free, flat and spreading, lower lobe smaller than the others, 2 stamens which spread laterally. - *Veronica*

- leaves all short-petiolate. - *Veronica beccabunga*
- leaves on the middle and upper portions of the flowering shoots sessile.
- capsule flattened, notched, wider than high, 5 to 9 seeds per locule, 1 to 2 mm long, leaves 4 to 20 times as long as wide. - *Veronica scutellata*
- capsule turgid, scarcely notched, barely wider than high, many, seeds less than 0.5 mm long, leaves 1.5 to 5 times as long as wide.
- leaves 1.5 to 3 times as long as wide, fruiting pedicels ascending, capsules higher than wide if not equal, flowers blue or violet. - *Veronica anagallis-aquatica*
- leaves 2.5 to 5 times as long as wide, fruiting pedicels spreading, capsules wider than high, flowers white to pink or pale blue. - *Veronica catenata*
- petal tube long or short but lobes not apparently free and all the same size, two, 4 or 5 erect stamens.
- stamen connective developed into a wide flap partially surrounding the anthers. - *Gratiola*
- pedicels with a pair of sepaloid bracteoles and 5 sepals at the summit. - *Gratiola neglecta*
- pedicels without bracteoles, only 5 sepals at the summit. - *Gratiola ebracteata*
- stamen connective not developed into a flap, sepal tube not 5-angled, no hairy patches on the abaxial petal lip.
- petal tube not 2-lipped, lobes equal or nearly so, leaves not finely dissected., with a petiole and a blade. - *Limosella aquatica*
- petal tube 2-lipped, lobes unequal, leaves finely dissected. - {*Limnophila sessiliflora*}

Key to the Umbelliferae of British Columbia

Key to the Zosteraceae of British Columbia

These are all marine plants.

- leaf sheaths deciduous, sometimes leaving a few scaly parts behind, leaf blades thin and translucent, rhizome with elongate internodes (1 to 3 cm long or more), with

2 thin roots at each internode, monoecious, spadix border projections inconspicuous, if present, usually established on sand or mud. - *Zostera*

- leaves 3-veined, 1 to 1.5 mm wide, sheaths split to the base, rare and probably introduced from Asia, known from Boundary Bay and Tsawwassen. - *Zostera japonica*
- leaves 5-veined or more, 1.5 to 12 mm wide, sheaths on sterile shoots are closed at the base, widespread native species. - *Zostera marina*
- basal portions of the leaf sheaths decay with age to bundles of fine, woolly fibres, leaf blades leathery, rhizome has short, thick internodes with 2 or more thick roots at each internode, dioecious, spadix bordered by conspicuous flap-like projections, usually established only on hard or rocky substrates. - *Phyllospadix*
- fertile stems branched, 4 to 12 dm long, spathes usually paired at the nodes. - *Phyllospadix torreyi*
- fertile stems unbranched, 0.5 to 4 dm long, spathes usually solitary at the nodes.
- leaves with 3, rarely 5, veins, margins entire. - *Phyllospadix scouleri*
- leaves with 5 or 7 veins, margins toothed towards the apex. - *Phyllospadix serrulatus*

Keys to the Aquatic Species within Genera

Key to the Alisma *of British Columbia*

- leaf blades lanceolate to ovate, scapes longer than the leaves (petiole plus blade), achenes centrally grooved at the top. - *Alisma plantago-aquatica*
- leaf blades narrowly lanceolate to linear, scapes shorter than the leaves (petiole plus blade), achenes 2-grooved at the tip. - Alisma gramineum

Key to the Azolla *of British Columbia*

A compound microscope is required to positively identify species in *Azolla*.

- glochidia with several (usually 3 or more) septa, leaves at least 0.7 mm long, plant 1 to 3 cm in diameter, submersed leaf-lobes much larger than the upper lobes, glochidia unbranched. - *Azolla mexicana*

- glochidia without any septa, leaves papillose, oblong to ovate, about 1 mm long, megaspores coarsely roughened, plants 1 to 10 cm long, branches numerous and open, lower submersed leaf-lobe about as large as the upper lobe, glochidia unbranched. - *Azolla filiculoides*
- glochidia with 1 or 2 apical septa, leaves nearly smooth, sub-orbicular, about 0.5 mm long, megaspores finely roughened, plants usually less than 3 cm long, few branches and crowded, lower submersed leaf-lobe glabrous, larger and paler than the upper lobe, glochidia may be branched. - *Azolla caroliniana*

Key to the Callitriche of British Columbia

Mature fruits are generally needed for positive identification of *Callitriche.*

- fruit encircled by a conspicuous wing-like margin, leaf bases joined by winged ridges. - *Callitriche stagnalis*
- fruit not winged or with only a narrow wing at the tip, leaf bases various.
- leaves all linear, 1-nerved, light green, leaf bases not joined by a wing, floral bracts absent, common species. - *Callitriche hermaphroditica*
- leaves various, upper often ovate and 3-nerved, bases joined by a wing-like ridge, floral bracts present.
- carpel face markings in regular vertical lines, fruit slightly wing-margined at the top and longer than broad. - *Callitriche verna*
- carpel face markings scattered, fruit not winged and as long as broad.
- fruits widest above the middle (obovate) leaves bidentate, midvein barely thickened at the end, emergent leaves may be over 5 mm wide, stems long. - *Callitriche heterophylla*
- fruits round or oblong, midvein thickened at the tip and protruding, leaves never over 5 mm wide, plants short and slender. - *Callitriche anceps*

Key to the Ceratophyllum *of British Columbia*

- leaf segments sub-capillary, mostly entire, delicate and light green, in deeper water and not surfacing, achene with 3 to 5 lateral spines on each side, not a 'weedy' species. - *Ceratophyllum echinatum*
- leaf segments capillary to linear and flattened, serrate to coarsely toothed, plant usually coarse and robust, dark green to almost black, usually surfacing, achene without lateral spines, 2 basal spines and 1 terminal spine only, a very 'weedy' species in eutrophic waters. - *Ceratophyllum demersum*

Key to the Chara *of British Columbia*

- stem not corticate, stipulodes in 1 tier, and alternating with the branchlets, branchlets tipped with a corona, no spine cells. - *Chara braunii*
- stems, and at least basal branchlet segment, corticate.
- spine cells in fascicles, axial cortex haplostichous, 1 corticate, stipulodes in two tiers, spiny looking, found in saline or brackish waters. - *Chara canescens*
- spine cells solitary or geminate, axial cortex diplostichous or triplostichous, two or 3 corticate.
- axial cortex diplostichous, stipulodes in 2 tiers, spine cells absent or small, rarely geminate, quite variable and widespread in hard-water lakes, smooth but marl encrusted. - *Chara vulgaris*
- axial cortex triplostichous, stipulodes usually in 2 tiers but may be in 1 tier, obscure or absent, spine cells usually absent or obscure but may be clustered, variable and widespread in hard-water and soft-water lakes, smooth appearance.

Chara globularis

Key to the Elodea *of British Columbia*

- leaves in whorls of 3.
- leaves rarely over 1.5 cm long.
- leaves 2 (1 to 4) mm wide, tapered abruptly to a blunt point, staminate flowers stalked and persistent. - *Elodea canadensis*

- leaves 1.5 (0.3 to 1.5) mm wide, tapered to a slender point, staminate flowers sessile and deciduous at anthesis. - *Elodea nuttallii*
- leaves (1.7) 2.0 to 2.6 cm long. - {*Elodea longivaginata*}
- leaves in the upper and middle part of the stem in pairs, lower ones irregular or alternate, (1.7) 2.0 to 2.6 cm long. - {*Elodea longivaginata*}

Key to the Aquatic **Equisetum** *of British Columbia*

This sub-set of all the *Equisetum* in BC share the following characteristics. Stems annual, usually with whorls of branches, cones blunt, fertile and sterile stems alike, ridges of the stem smooth or without tubercles or spicules, often cross-wrinkled, cones appear in summer.

- central cavity well over 1/2 the diameter of the stem, teeth may be deciduous, 10 to 40 ridges, teeth persistent, black but not hyaline margined, stomates in 1 row and not sunken. - *Equisetum fluviatile*
- central cavity less than 1/3 the diameter of the stem, teeth not deciduous, black and hyaline margined, 5 to 10 ridges. - *Equisetum palustre*

Key to the **Gratiola** *of British Columbia*

- there are 2 sepaloid bracts at the apex of the pedicel and thus apparently 7 sepals. - *Gratiola neglecta*
- pedicels without bracts and sepals thus evidently only 5. - *Gratiola ebracteata*

Key to the **Hydrocotyle** *of British Columbia*

- leaves orbicular to reniform, crenate or shallowly lobed, centrally peltate. - *Hydrocotyle verticillata*
- leaves 5 to 6 lobed nearly to the centre, not peltate. - *Hydrocotyle ranunculoides*

Key to the **Isoetes** *of British Columbia*

A compound or dissecting microscope is required to positively identify species in *Isoetes*. Magnification over 10x is needed to examine the megaspore surface.

- corms 3-lobed, plants primarily terrestrial, grow in grassy, ephemeral pools which are wet in winter.

- corms 2-lobed, plants primarily of lake bottoms but sometimes exposed in summer due to water level fluctuations.
- megaspores spiny.
- megaspores vary greatly in size and are often aborted, spines blunt and dense. (this is probably a hybrid of *Isoetes echinospora* and *Isoetes maritima*). - *Isoetes truncata*
- megaspores uniform in size, spines sharp, not crowded.
- plants flaccid, megaspore spines elongate with pointed tips, uniform in size, microspores smooth. - *Isoetes echinospora*
- plants rigid, megaspore spines stubby and blunt, shorter and denser along the ridges, microspores spiny. - *Isoetes maritima*
- megaspores smooth or with ridges but not spiny.
- lowermost leaves in 2 ranks, megaspores over 0.5 mm in diameter, microspores rugose and over 36μm long. - *Isoetes occidentalis*
- lowermost leaves spirally arranged, megaspores less than 0.5 mm in diameter, microspore spiny and less than 30μm long.
- base of the leaves blackened, hyaline wing margins of the leaves extending 1 to 5 cm above the sporangium, sporangia brown spotted. - *Isoetes howellii*
- base of the leaves green, hyaline wing margins of the leaves not extending more than 1 cm above the sporangium, sporangia without colour. - *Isoetes bolanderi*

Key to the Lemna of British Columbia

- fronds oblong to lanceolate, 6-12 mm long and connected in small groups by stalks of the same length, matted, generally submersed, colonies. *Lemna trisulca*
- fronds oval to round, less than 6 mm long, no stalks, solitary or in small attached groups, floating on the surface, usually 1 nerved, 2 to 4 mm long. - *Lemna minor*

Key to the Limnobium of British Columbia

- floating leaves long-petiolate, leaves flat on the top with a thin spongy layer beneath and red-brown markings on the upper surface. - *Limnobium spongia*
- floating leaves short-petiolate, leaves arched on the top with a thick spongy layer beneath and green with no coloured markings on the upper surface. - *Limnobium laevigatum*

Key to the Aquatic Myosotis of British Columbia

- corolla limb 2 to 5 mm wide, style shorter than the nutlets, not stoloniferous, often recumbent. - *Myosotis laxa*
- corolla limb 5 to 10 mm wide, style as long as or longer than the nutlets, stoloniferous, mostly erect and not creeping or recumbent. - *Myosotis scorpioides*

Key to the Myriophyllum of British Columbia

Submersed leaves are simply pinnate. The key includes both native and two introduced species, one of which is widespread in aquaria and outdoor garden pools.

- flowers in the axils of cauline submersed leaves.
- leaves in whorls of 3 to 4, or scattered, fewer than 10 leaf segments on each side of the rachis, monoecious, 4 stamens, plants often reddish, fully submersed. - *Myriophyllum farwellii*
- leaves in whorls of 4 to 6, more than 10 leaf segments on each side of the rachis, dioecious, 8 stamens, plants a pallid yellowish-green, apical portion of stem often sprawled over the surface of the water or the adjacent shore. - {*Myriophyllum aquaticum*}
- flowers in the axils of bracts on emergent, terminal spikes.
- four stamens, 4 to 6 leaves per whorl, bracts conspicuous.
- floral bracts delicate, deeply incised to serrate, spike short and delicate, widespread in sloughs of the lower Fraser Valley. - *Myriophyllum hippuroides* and *Myriophyllum pinnatum*

- floral bracts ovate and toothed, spike long, robust and inflated, introduced in several park and garden ponds of south-western British Columbia. - {*Myriophyllum heterophyllum*}
- eight stamens, 3 to 5 leaves per whorl, floral bracts various.
- floral bracts smaller than the flowers, inconspicuous, and nearly entire (the lowest few may be larger and pinnate but the upper ones are small), leaf whorls in the central portion of the stem are over 1 cm apart and not crowded, monoecious.
- no turions, rhizomatous, 10 to 16 leaf divisions less than 2 mm apart, leaves make right or obtuse angles with the stem, leaf tips 'squared', all leaf segments straight and all of nearly the same length. - *Myriophyllum spicatum*
- turions present, not rhizomatous, 6 to 12 leaf segments over 2.5 mm apart, leaves make acute angles with the stem, leaf tips 'acute', basal leaf segments curved and much longer than the apical. - *Myriophyllum sibiricum* [*Myriophyllum exalbescens*]
- floral bracts usually longer than the flowers and rarely entire, the leaf whorl spacing varies, monoecious or dioecious.
- dioecious (male and female flowers on separate small plants) found only on exposed mud banks when the water level drops in summer, female bracts and leaves entire to scarcely and irregularly divided, male bracts and leaves entire to pectinate-pinnate, submersed leaves scattered and irregular. - *Myriophyllum ussuriense*
- dioecious or monoecious, flowering plants found in water, bracts large and conspicuous, shape variable.
- floral bracts pinnate to pectinate, greenish, leaves often crowded on the stem and delicate, usually more than 10 leaf divisions. - *Myriophyllum verticillatum*
- floral bracts pectinately parted below becoming dentate in the middle and almost entire above, reddish, leaves well spaced on the stem and robust, generally.

Key to the Nitella *of British Columbia*

- the end segment of the branchlets is one-celled.
- heteroclemous, 2 kinds of branchlets in a whorl, 1-forked alternating with one-celled, bright green, compact, branchlets swollen. - *Nitella clavata*
- homeoclemous, all branchlets in a whorl the same, generally 1-forked.
- fertile heads small and densely compact, no mucous, generally no terminal dactyls on the sterile branchlets. - *Nitella acuminata*
- fertile heads absent or loose and not compact.
- dactyl apices not long acuminate, may be acute, blunt or apiculate, (a small variety may have 2-forked branchlets occasionally). - *Nitella flexilis*
- dactyl apices are long acuminate, gametangia on branchlets, generally not marl covered. - *Nitella acuminata*
- the end segment of the branchlets is 2-or-more-celled, the end cells are very much smaller than the penultimate cell, branchlets furcate, heads terminal, branchlets homeoclemous.
- robust plants, over 15 cm high, axes over 600 microns in diameter, dactyls may be small. - *Nitella furcata*
- smaller plants, seldom over 15 cm tall, axes 200 to 450 microns in diameter, dactyls uniform.
- lowest branchlet node fertile, dactyls 2- to 3-celled, heads may have mucous, oospore membrane granular or felt-like. - *Nitella gracilis*
- lowest branchlet node sterile, dactyls 2-celled, heads without mucous, oospore membrane reticulate. - *Nitella tenuissima*

Key to the Nuphar *of British Columbia*

In *Nuphar* the large, showy, yellow, perianth members are sepals; the petals are smaller than the stamens and inconspicuous.

- sepals 6 to 8 (usually 6), 2.5 to 3.5 cm long, stamens yellow. - *Nuphar variegatum*

- sepals 8 to 17 (usually 9), (3) 3.5 to 6 cm long, stamens reddish. - *Nuphar polysepalum*

Key to the Nymphaea *of British Columbia*

- flowers yellow, 6 to 13 cm in diameter, leaves dark green and blotched above, brownish with black dots below, plant tuberous. - *Nymphaea mexicana*
- flowers white pink or red, leaves not blotched.
- small plant with leaves up to 8 cm wide and flowers up to 5 cm wide, petals seven to 15, stigmas 6 to 9, flowers white. - *Nymphaea tetragona*
- larger plants with leaves over 8 cm wide and flowers over 5 cm wide, petals and stigmas numerous, flowers white, pink, or rose.
- flowers white or tinged with pink, 20 to 32 petals, leaves scattered on the rhizome, flowers strongly scented, leaves usually purplish beneath. - *Nymphaea odorata*
- flowers white, 12 to 24 petals, leaves crowded on the rhizome, flowers not strongly scented, leaves greenish beneath. - *Nymphaea alba*

Key to the Nymphoides *of British Columbia*

- flowers yellow, axillary, no cluster of roots on the petiole, leaves primarily from branching stems. - *Nymphoides peltata*
- flowers white, in clusters on the petioles, a cluster of roots on the petiole, leaves mostly basal with long slender petioles.
- leaf blades ovate, mostly under 5 cm long, seeds smooth, leaves not dark-punctate beneath. - *Nymphoides cordatum*
- leaf blades orbicular, 8 to 15 cm in diameter, seeds glandular, warty, leaves dark-punctate or pitted beneath. - *Nymphoides aquaticum*

Key to the Phyllospadix *of British Columbia*

- fertile stems branched, 4 to 12 dm long, spathes usually paired at the nodes. - *Phyllospadix torreyi*

- fertile stems unbranched, 0.5 to 4 dm long, spathes usually solitary at the nodes.
- leaves with 3, rarely 5, veins, margins entire. - *Phyllospadix scouleri*
- leaves with 5 or 7 veins, margins toothed towards the apex. - *Phyllospadix serrulatus*

Key to the Aquatic Polygonum *of British Columbia*

All our BC aquatic *Polygonum* share the following characteristics. Stipules red or brown, cylindrical or funnel-like, sometimes bristly, usually perennial and rhizomatous, fresh water.

- flowers rose in 1 or 2 terminal inflorescences, perianth not glandular, achenes lenticular. - *Polygonum amphibium*
- flowers whitish, pinkish or greenish-white in 2-to-many contracted panicles, perianth may be glandular, achenes mostly triquetrous.
- perianths glandular-punctate, glands sessile, stamens 6 or 8, achenes brown, glandular and dull, perianth usually 4-lobed, stamens usually 6. - *Polygonum hydropiper*
- perianth not glandular-punctate, achenes dark brown or black, stamens 8.
- inflorescence ending in spicate, slender, interrupted racemes mostly over three cm long, nerves of the perianth segments not branched and recurved, stamens 8, perianth 5-lobed. - *Polygonum hydropiperoides*
- inflorescence of several short, thick, continuous racemes, rarely over three cm long, nerves of the perianth segments branched and recurved, perianth 4 to 5 lobed. - *Polygonum lapathifolium*

Key to the Potamogeton *of British Columbia*

- stipules forming a sheath around the stem partly below the base of the leaf blade and partly above, the leaf attached 1 cm or more above the node.
- leaves 3 to 4 mm wide, flat, rough on the edges, stiff, with a broad mid-vein and over 20 lateral veins.

- leaves less than 3 mm wide, cylindrical, smooth, soft, with only one inconspicuous vein.
- leaves acute and sharp pointed at the apex, fruits 2.5 to 4 mm long with a short beak, the surface of fruits without radial striation (at 10x). - *Potamogeton pectinatus*
- leaves obtuse and blunt at the apex, fruits 2 to 3 mm long, beakless, surface of the fruits with radial striation (at 10x).
- primary stems 0.5 to 1 mm in diameter, all leaves filiform, 0.2 to 0.5 mm wide with tight sheaths, spikes with 2 to 5 whorls of flowers, fruits 2 to 2.5 mm long. - *Potamogeton filiformis*
- primary stems 1 to 3 mm in diameter, lower main stem leaves with blades 1 to 2 mm wide and with loose inflated sheaths, upper branch leaves filiform, spikes with 5 to 12 whorls of flowers, fruits 2.5 to 2.8 mm long. *Potamogeton vaginatus*
- stipules forming a sheath around the stem only above the base of the leaf blade, the leaf attached at the node.
- submersed leaves linear, ribbon-like or cylindrical, less than 5 mm wide, parallel margins.
- submersed leaves cylindrical, terete.
- floating leaves 2.5 to 6 cm wide, usually cordate, large ones with (18) 21 to 35 nerves, submersed leaves arise from the main stem, mature fruit 3.7 to 4.5 mm long including the beak, obscurely keeled, interlacunar bundles in several circles throughout the aerenchyma. - *Potamogeton natans*
- floating leaves 1 to 3 cm wide, usually rounded or cuneate at the base, largest ones with 11 to 19 nerves, submersed leaves arise from the branches of the main stem, mature fruit, including the beak, 3 to 3.4 mm long, interlacunar bundles develop only in the outer circle of the aerenchyma. - *Potamogeton oakesianus*
- submersed leaves flat.
- submersed leaves with 9 to 35 veins. - *Potamogeton zosteriformis*

- submersed leaves with 1 to 7 veins.
- stipules strongly fibrous, becoming whitish, especially on the turions, base of turions strongly ribbed.
- leaves obtuse or rounded and slightly mucronate, not conspicuously 2 ranked,, blades thin, 1.5 to 3.5 mm wide, 5 to 7 veins, a narrow cellular-reticulate band along the midrib, turions fan-shaped, peduncle flattened. - *Potamogeton friesii*
- leaves gradually tapered into a sharp bristle tip, conspicuously 2 ranked, blades firm, 0.5 to 2.5 mm wide, convolute with 3 (5) veins, no cellular-reticulate band along the midrib, turions slender, peduncles terete. - *Potamogeton strictifolius*
- stipules delicate, not fibrous, greenish or brownish, base of the turions smooth.
- leaves (2) 3 to 4 mm wide, rounded at the apex, fruits (3) 3.5 to 4 mm long. - *Potamogeton obtusifolius*
- leaves 0.3 to 3 mm wide, acute to obtuse or mucronate, fruits 1.8 to 2.8 mm long
- fruits with a distinct dorsal keel, veins on the stipules evident as ridges running the full length of the stipule, glands at the base of the stipules either lacking or poorly developed. - *Potamogeton foliosus*
- fruits without a dorsal keel, veins on the stipules obscure and faint, glands at the base of the stipules usually well developed.
- stipules connate, fused along the stem, at least when young, mature fruits broadest above the middle, plants sparsely branched. - *Potamogeton pusillus*
- stipules convolute, mature fruits broadest below the middle, plants well branched. - *Potamogeton berchtoldii*
- submersed leaves lanceolate or ovate, more than 5 mm wide.
- leaf margins serrate, beak of the fruit as long as the fruit body or longer. - *Potamogeton crispus*
- leaf margins entire, beak of the fruit much shorter than the fruit body.

- submersed leaves ribbon-like with parallel side, 5 to 10 mm wide, limp and flaccid, median band of lacunae several cells wide at least 1/4 the width of the blade, stems compressed. - *Potamogeton epihydrus*
- submersed leaves lanceolate to ovate with sides not parallel, either no median cellular-reticulate band or the band less than 1/4 the width of the blade, stems terete.
- submersed leaves sessile, cordate or rounded at the base and clasping the stem.
- leaves ovate-oblong, mostly 10 to 20 cm long, with a cucullate apex, fruits more than 4 mm long, stems with many interlacunar vascular bundles. - *Potamogeton praelongus*
- leaves rounded, ovate or elongate-ovate, 1 to 10 cm long, no cucullate apex, fruits under 3.5 mm long, no interlacunar vascular bundles.
- stipules coarse disintegrating into persistent whitish fibres, peduncles clavate, 1.5 to 25 cm long, fruits with a cavity in the endocarp loop. - *Potamogeton richardsonii*
- stipules delicate, lacking on old specimens, peduncles not clavate, 1 to 9 cm long, fruits without a cavity in the endocarp loop. - *Potamogeton perfoliatus*
- submersed leaves petiolate or sessile, not clasping the stems.
- submersed leaves 2 to 5 cm wide, tapering into distinct petioles, stipules over 3 cm long.
- floating leaves cordate, submersed leaves folded and strongly falcate, more than 20 veins, interlacunar bundles well developed throughout. - *Potamogeton amplifolius*
- 20. floating leaves rounded or cuneate at the base, submersed leaves flat and not falcate, with fewer than 20 veins, interlacunar bundles developed in only one circle or absent.
- petioles of submersed leaves shorter than 2 cm, interlacunar bundles well developed forming one circle, endodermis of U-cells. - *Potamogeton illinoensis*

- petioles of submersed leaves longer than 2 cm, interlacunar bundles absent, endodermis of O-cells. - *Potamogeton natans*
- submersed leaves mostly less than 2 cm wide, sessile, or with petioles under 0.5 cm long, stipules under 3 cm long.
- plants with a reddish tinge, usually not branched, floating leaves, if present, not markedly different from the submersed leaves, no interlacunar bundles, endodermis of O-cells. - *Potamogeton alpinus*
- plants greenish, freely branched, floating leaves markedly different from the submersed leaves, interlacunar bundles well developed in one circle, endodermis of U-cells. - *Potamogeton gramineus*

Key to the Aquatic Ranunculus of British Columbia

- flowers white, submersed leaves dissected into ultimately filiform segments.
- two to 7 glabrous, beakless achenes, plants and receptacle glabrous, pedicels in the axils of ternately lobed floating leaves, leaves 2 to 3 times divided into 8 to 12 segments. - *Ranunculus lobbii*
- ten to 80 glabrous or hirsute achenes, plants and receptacles hirsute.
- thirty to 80 achenes, flowers may have yellow bases, leaf blades sessile on the stipular base. - *Ranunculus circinatus*
- ten to 25 achenes, flowers all white, leaf blades petiolate. - *Ranunculus aquatilis*
- flowers yellow, submersed leaves, if present, simple to ternately divided.
- leaves simple and entire to serrate, may be shallowly lobed, achenes glabrous.
- achenes 50 to 200 in a columnar head, longitudinally ribbed. - *Ranunculus cymbalaria*
- achenes 5 to 60 in a globular head, not longitudinally ribbed, stems decumbent to prostrate, rooting at the nodes, nectary scales usually broader than long.

- leaves all, or in part, lobed, parted or ternately dissected to filiform segments, achenes pubescent.
- annuals, erect, not nodally rooting, achene beaks inconspicuous. - *Ranunculus sceleratus*
- perennials, floating or reclining, nodally rooting, achene beaks conspicuous.
- leaves deeply lobed or parted, achenes not corky margined.
- leaves deeply 3 parted with narrow lobes distally acute, achene beaks one quarter the length of the achenes, receptacles hirsute, 5 sepals and petals. - *Ranunculus gmelinii*
- leaves 3 lobed, distally rounded, achene beaks 1/10 the length of the achenes, receptacles glabrous, 3 sepals and petals. - *Ranunculus hyperboreus*
- leaves, at least the submersed ones, 3 to 5 times ternately dissected into filiform segments less than 2 mm wide, achenes corky margined. - *Ranunculus flabellaris*

Key to the Sagittaria of British Columbia

In *Sagittaria* the usual non-achene key characters of bract length and shape and pedicel lengths have been found to be inconsistent and sufficiently variable even within one population as to be virtually useless as key characters. Many flowering, but non-fruiting, collections are difficult to identify.

- leaves basally sagittate or hastate, stamens 7 to 25, achenes densely packed on the receptacle, plants emergent in shallow water. - *Sagittaria*
- monoecious, mature achenes 2.0 to 2.5 mm long with a beak less than 0.5 mm long pointing forward from the tip of the achene. - *Sagittaria cuneata*
- monoecious or dioecious, mature achenes (2.5) 3.0 to 3.5 (4.0) mm long with a beak about 1 mm long at right angles to the body of the achene. - *Sagittaria latifolia*
- leaves gradually tapered or cordate at the base; broadly ribbon-like, submersed in deep water or floating on the surface, ribbon-like or tapered at the base and oval, flaccid, plants generally sterile. - *Sagittaria*

Key to {Salvinia}

There is still considerable taxonomic confusion in *Salvinia* and since these plants are only expected as sporadic, non-persistent escapes from aquaria and garden ponds, there is little benefit in trying to identify individual species. The genus is readily recognizable but individual species are often only distinguished on technical and difficult characters. There is hybridization occurring and some 'species' are sterile hybrids and serious tropical weeds.

- the four hairs on the tips of each leaf papilla are joined at the distal end. - {*Salvinia auriculata*} complex [*Salvinia molesta*]
- the hairs on the tips of the leaf papillae are free. - other {*Salvinia*} species

The *Salvinia auriculata* complex includes, in part, such species as {*Salvinia herzogii*}, {*Salvinia biloba*}, and {*Salvinia auriculata*}

The other species include, in part, such species as {*Salvinia natans*}, {*Salvinia sprucei*}, {*Salvinia cucullata*}, {*Salvinia rotundifolia*}, {*Salvinia minima*} and {*Salvinia oblongifolia*}.

Key to the Sparganium of British Columbia

- two (1) stigmas over 2 mm long, the achenes truncate-pyriform and narrowed abruptly to the beak, inflorescence usually branched, staminate heads above the pistillate heads, large robust plants. - *Sparganium eurycarpum*
- one stigma, achenes fusiform and tapering to the beak, inflorescence usually simple, staminate heads below the pistillate heads.
- achene beaks less than 1.5 mm long or beakless, staminate head solitary.
- achene beaks less than 0.5 mm long to absent, staminate head closely adjacent to the upper pistillate head, basal leaves (0.5) 1 to 3 (5) mm wide, opaque, often yellow. - *Sparganium hyperboreum*
- achene beaks 0.5 to 1.5 mm long, staminate head distant from the upper pistillate head, basal leaves (1.5) 2 to

6 (10) mm wide, translucent, usually dark green. - *Sparganium natans*

- achene beaks 1.5 to 5 mm long, there may be more than 1 staminate head.
- achene beaks conspicuously curved. - *Sparganium fluctuans*
- achene beaks straight or slightly curved.
- staminate head usually solitary, closely adjacent to the upper pistillate head. - *Sparganium glomeratum*
- staminate heads (1) 3 to 8, distant from the upper pistillate head.
- pistillate heads (1) 2 to 3 (4), usually crowded and appearing as one elongated head terminating the inflorescence, achene beaks 1.5 to 2.0 (2.2) mm long. - *Sparganium angustifolium*
- pistillate heads (3) 4 to 7 (10), distant and distinct, achene beaks 2 to 4.5 (6) mm long. - *Sparganium emersum*

Key to the Utricularia *of British Columbia*

- leaves divided into ultimately terete or threadlike segments.
- leaves divided into fewer than 5 final threadlike segments, leaf margins glabrous, bladders scarce on a small delicate plant usually floating at the surface or entangled in other rooted plants. - *Utricularia gibba*
- leaves 'pinnatifid', more than 20 terete final segments, hairy leaf margins, many bladders on the ordinary leaves, a robust plant to several meters long, usually lying on the sediment surface. - *Utricularia vulgaris*
- leaves divided di- or tri-chotomously into ultimately flattened segments.
- ordinary leaves generally with a few bladders, leaf margins glabrous, the terminal leaf segments are acuminate. - *Utricularia minor*
- ordinary leaves rarely, if ever, with bladders which are found on separate subterranean branches, leaf margins

hairy, the ultimate leaf segment are awned. - *Utricularia intermedia*

Key to the Aquatic Veronica of British Columbia

- leaves all short-petiolate. - *Veronica beccabunga*
- leaves, at least those on the middle and upper portions of the flowering shoots, sessile.
- capsule flattened, notched, wider than high, 5 to 9 seeds per locule, 1 to 2 mm long, leaves 4 to 20 times as long as wide. - *Veronica scutellata*
- capsule turgid, scarcely notched, barely wider than high, many seeds less than 0.5 mm long, leaves 1.5 to 5 times as long as wide.
- leaves 1.5 to 3 times as long as wide, fruiting pedicels ascending, capsules higher than wide if not equal, flowers blue or violet. - *Veronica anagallis-aquatica*
- leaves 2.5 to 5 times as long as wide, fruiting pedicels spreading, capsules wider than high, flowers white to pink or pale blue. - *Veronica catenata*

Key to the Wolffia of British Columbia

- plants floating just below the surface, sub-globose, upper surface rounded, green, not puncticulate, 0.5-1.0 mm long. - *Wolffia columbiana*
- plants floating on the surface, ellipsoidal, upper surface flattish, white- or brown-punctilate, 0.5-1.2 mm long and about 1/2 as wide. - *Wolffia borealis*

Key to the Zostera of British Columbia

- leaves 3-veined, 1 to 1.5 mm wide, sheaths split to the base, rare and probably introduced from Asia, known only from Boundary Bay and Tsawwassen. - *Zostera japonica*
- leaves 5-veined or more, 1.5 to 12 mm wide, sheaths on sterile shoots are closed at the base, widespread native species. - *Zostera marina*

Distribution, Selected Common Names and Habitat within BC

Not all of the plants in these keys are native or naturalized in British Columbia. Some are garden or aquarium plants that are introduced from time to time but do not persist, or are in cultivation but do not survive long when they escape. Others are major invasive weeds in other parts of the world and, although they have not yet been found in British Columbia, they are found in neighbouring areas and are expected to be found in British Columbia eventually. People deliberately, and inadvertently, introduce many aquatic plants, some of which do eventually become permanently established.

The following brief notes give an indication of the habitats of native and introduced aquatic and wetland plants in British Columbia and their general distribution range. A few selected common names are also given with the preferred British Columbia common name listed first and other common names following in alphabetical order. There are literally dozens of local common names, in many languages, for some of the more widespread species of aquatic weeds. A more complete list of these common names can be found in Aquatic Plants of British Columbia: Common Names, Selected References, Synonymy and Classification by Life-Forms and Habitat. Occasionally other pertinent or interesting notes are included.

Alisma: The species in this genus are commonly known as Mud plantains, Plantains and Water plantains

Alisma Gramineum J. G. Gmel: Narrow leaved water plantain, Grass leaved water plantain Shallow water of lakes, rivers, marshes and tidal flats in the lowland, steppe and montane zones; uncommon in southern BC; circumpolar.

Alisma Plantago-aquatica L: American water plantain, Arrowhead, Broadleaf water plantain, Common water plantain, Heart shaped water plantain, Mud plantain Shallow water of ponds, marshes and ditches in the lowland, steppe and montane zones; common in southern BC south of 52 degrees N; rare northward.

Azolla: The species in this genus are commonly known as Fairy mosses, Mosquito ferns, Water velvets and Water ferns

Azolla Caroliniana Willd: Carolina water fern Surface of water in sloughs and ditches in the lowland zone; rare in south-western BC, primarily in the lower Fraser Valley. Introduced and available in garden shops, aquatic plant nurseries and the aquarium trade.

Azolla Filiculoides Lam: Large mosquito fern, Duckweed fern, Pacific azolla Surface of water in sloughs and ditches in the lowland zone; rare in south-western BC, primarily in the lower Fraser Valley. A European introduction with large populations where found. Commonly available in garden shops, aquatic plant nurseries and the aquarium trade.

Azolla Mexicana K. B. Presl: Mexican mosquito fern Surface of the water in sloughs and pools in the montane zone; rare in south-central BC, known from the Salmon Arm and Sicamous area.

Brasenia Schreberi Gmel: Watershield, Dollar bonnet, Frog leaf, Junsai, Little water lily, Purple bonnet, Purple wendock, Schreber watershield, Water target Ponds, lakes and slow-moving streams in the lowland, steppe and montane zones; common in south-western BC, less frequent in south-central BC.

{Cabomba Caroliniana Gray}: Fanwort, Carolina water shield, Fish grass, Parrot feather, Washington grass A commonly introduced aquarium and garden pond plant, sometimes dumped but not known to overwinter outdoors in BC, established in Oregon and the lower Columbia River of Washington.

Calla Palustris L: Wild calla, Bog arum, Calla lily, Water arum

Shallow water of swamps, marshes and lakes in the montane zone; frequent in BC north of 52 degrees N and east of the Coast Mountains, rare in south-central BC; circumpolar.

***Callitriche*:** The species of this genus are commonly known as Starworts, Water chickweeds and Water starworts.

***Callitriche Anceps* Fern:** Two edged water starwort Shallow water of ponds and lakes in the lowland and montane zones; rare along the coast in BC.

***Callitriche hermaphroditica* L:** Northern water starwort, Autumnal starwort Shallow water of sow streams, ditches and sloughs in the lowland, steppe and montane zones; common in south-western BC, less frequent in south-central and south-eastern BC; circumboreal.

***Callitriche Heterophylla* Pursh:** Diverse leaved water starwort, Large water starwort Shallow water of slow streams, ditches, lakes and ponds in the lowland zone; frequent on southern Vancouver Island, rare in south-eastern BC.

Callitriche Stagnalis Scop: Pond water starwort, Common starwort, Common water starwort Shallow water of slow streams, ditches and ponds; infrequent in southern BC; introduced from Europe.

Callitriche Verna L: Spring water starwort, Common water starwort, Vernal water starwort Shallow water of slow streams, ditches and shorelines in the lowland, steppe and montane zones; frequent in southern BC; circumboreal.

Caltha Natans Pallas: Floating marsh marigold Shallow water of ponds and lakes in the montane zone; locally frequent in north-eastern BC; amphiberingian.

Caltha Palustris L: Yellow marsh marigold, Cowslip Wet sites, bogs and shallow water in the lowland zone; rare along the coast.

Ceratophyllum: The species in this genus are commonly known as Coontails, Foxtails, Hornworts and Water mosses.

***Ceratophyllum Demersum* L:** Common coontail, Common hornwort, Grey pimpled hornwort, Morass weed Ponds, lakes and slow streams in the lowland, steppe and montane zones; common in southern BC, less frequent northwards.

***Ceratophyllum Echinatum* Gray:** Spring hornwort, Prickly coontail, Spiny hornwort Lakes and sloughs in the lowland and montane zones; locally frequent on southern Vancouver Island and the Gulf Islands, rare in the lower Fraser Valley and along the Alaska Highway.

***Chara*:** The species of this genus are commonly known as Chara, Musk grasses, Skunk grasses and Stoneworts.

***Chara Braunii* Gm:** Little data is available on the abundance and distribution of charophytes in BC.

***Chara Canescens* Desv. and Lois:** Little data is available on the abundance and distribution of charophytes in BC.

***Chara Globularis* Thuill:** Little data is available on the abundance and distribution of charophytes in BC.

Chara Vulgaris L: Little data is available on the abundance and distribution of charophytes in BC.

Crassula Aquatica (L.) Schoenl: Pygmyweed Vernal pools and mud flats in the lowland zone; rare, scattered throughout southern BC; circumpolar. [Tillaea aquatica L.]

Dulichium Arundinaceum (L.) Britt: Three way sedge, Dwarf bamboo, Pond sedge Wet meadows, shallow water of lakes and stream sides in the lowland, steppe and montane zones; frequent in south-western BC, rare in south-central BC.

{Egeria Densa Planch.}: Brazilian waterweed, Anacharis, Argentine water weed, Brazilian elodea, Dense water weed, Giant elodea, Leafy elodea, Oxygen weed, South american elodea Lakes, ponds and streams in the lowland zone; known from only a few sites in Victoria and Vancouver; introduced from South America. Naturalized and prolific where present and of concern as a weed displacing native species. Readily available in garden centres and aquarium shops. Well established in over a dozen lakes in Washington State.

{Eichhornia Crassipes (Mart.) Solms}: Water hyacinth, Florida devil An often introduced garden pond species, but not yet known to overwinter outdoors in BC. Readily available in garden centres and aquarium shops. (None of the authors specimens, which grow and multiply prolifically in the greenhouse and outdoors in summer, has ever survived outdoors over the winter in Victoria).

Elatine Rubella Rydb: Three flowered waterwort Ditches, mud flats, shallow ponds and lake shores in the lowland zone; rare in south-western BC.

Elodea: The species of this genus are commonly known as Ditch mosses, and Waterweeds.

Elodea Canadensis Rich: Canadian waterweed, American elodea, Canadian pondweed, Common elodea, Water thyme Lakes, ponds, streams and ditches in the lowland, steppe and montane zones; common in south-western BC, rare northward.

Elodea Longivaginata St. John: Long leaved waterweed This species has not been confirmed in BC yet but is found east of the Rockies in Alberta, Montana and Wyoming. It is included in the keys in case a specimen is found in south-eastern BC or is introduced with garden plants. This species should be able to establish and overwinter successfully if introduced

Elodea Nuttallii (Planch.) St. John: Nuttall's waterweed, Slender waterweed, Western waterweed Lakes, ponds and streams in the lowland, steppe and montane zones; rare in southern BC.

***Equisetum*:** The species of this genus are commonly called horsetails or scouring rushes.

Equisetum Fluviatile L: Swamp horsetail, Water horsetail Shallow water at lake margins in the lowland, steppe and montane zones; frequent throughout BC; circumpolar. This species often forms very extensive colonies with a distinctive horizontally-banded appearance en masse.

Equisetum Palustre L: Marsh horsetail Shallow water of marshes and swamps, stream banks and forests in the lowland, steppe and montane zones; frequent in southern BC, infrequent elsewhere; circumpolar. Commonly found marginally in subalpine lakes throughout BC.

***Gratiola*:** The species of this genus are commonly known as hedge hyssops.

Gratiola Ebracteata Bentham: Bractless hedge hyssop Wet sites and shallow marginal water in the lowland zone; frequent in south-western BC.

Gratiola neglecta Torr: Common american hedge hyssop, Clammy hedge hyssop, Obscure hedge hyssop Wet sites and shallow marginal water in the lowland, steppe and lower montane zones; rare in southern BC.

Heteranthera Dubia (Jacq.) Macmill: Water star grass, Mud plantain, Water star weed Ponds, lakes and slow-moving streams in the steppe and montane zones; infrequent in south-western and south-central BC.

Hippuris Vulgaris L: Common mare's tail, Four leaved mare's tail, Water mare's tail Shorelines, lakes, ponds and quiet open water in the lowland, steppe and montane zones; common throughout BC, less frequent along the coast; circumpolar.

{Hydrilla Verticillata (L.) Royle}: Hydrilla, Florida elodea, Oxygen grass, Oxygen weed, Water thyme Lakes, ponds streams and other permanent water. This plant has not been found in BC yet, but it is established in Washington State. It is expected to overwinter successfully and become a weed once it is introduced.

Hydrocotyle Ranunculoides L. f: Floating water pennywort Ponds, shallow water of marshes and wet sites in the lowland zone; rare on south-eastern Vancouver Island.

Hydrocotyle Verticillata Thunb: Whorled water pennywort Streams, shallow water of marshes and wet sites in the lowland zone; rare, known from the lower Fraser valley.

Isoetes: The species of this genus are commonly known as Quillworts, Bracksen grasses, Merllyn's grasses or Octopus plants.

Isoetes Bolanderi Engelm: Bolander's quillwort Shallow water of lakes and muddy sites in the subalpine zone; rare in south-eastern BC, known only from Akamina Pass.

Isoetes Echinospora Dur: Bristle like quillwort Lakes in the lowland to subalpine zones; infrequent throughout BC; circumpolar.

Isoetes Howellii Engelm: Howell's quillwort Shallow water of lakes (exposed in summer) in the steppe and montane zones; rare in south-central BC, known from Kamloops, Mara and Shuswap Lakes.

Isoetes Maritima Underw: Coastal quillwort Shallow water of lakes and other shallow water habitats in the lowland and montane zones; frequent in coastal BC, rare in south-central BC; amphiberingian.

Isoetes Nuttallii A. Br: Nuttall's quillwort Vernal pools and ephemeral winter seepages in the lowland zone; rare on south-eastern Vancouver Island.

Isoetes Occidentalis Henderson: Western quillwort Lakes in the lowland, steppe and montane zones; frequent in southern BC.

Isoetes Truncata (A. A. Eaton) Clute: Slashed quillwort Shallow water of lakes (exposed in summer) in the lowland zone; infrequent on south-eastern Vancouver Island (may be a hybrid between Isoetes echinospora and Isoetes maritima).

Juncus Supiniformis Engelm: Spreading rush Shallow water, wet muck, lake shores and open bogs in the lowland and montane zones; common in BC west of the Coast-Cascade Mountains.

Lagarosiphon Major Ridley: Lagarosiphon, African elodea, Oxygen weed This 'weedy' species is distributed as an aquarium plant but it has not yet been found out of cultivation in BC.

***Lemna*:** The species of this genus are commonly known as Duckmeats, Duckweeds and water lentils.

Lemna Minor L: Common duckweed, Lesser duckweed, Small duckweed, Water lentil Ponds, lakes and slow moving streams in the lowland, steppe and montane zones; common throughout BC south of 55 degrees N, less frequent northward, absent in north-western BC and the Queen Charlotte Islands; circumpolar.

Lemna Trisulca L: Star duckweed, Chain of stars, Forked duckweed, Ivy leaved duckweed Ponds, lakes and slow moving streams in the lowland, steppe and montane zones; frequent throughout BC, absent in the Queen Charlotte Islands; circumpolar.

Lilaea Scilloides (Poir.) Haum: Flowering quillwort Mud flats, ponds, shallow water of lakes and wet sites in the lowland zone; rare in coastal BC, bipolar disjunct.

Lilaeopsis Occidentalis Coult. and Rose:Western lilaeopsis Shallow water of marshes, lakes and tidal shores in the lowland zone; infrequent along the coast.

{Limnophila Sessiliflora Blume}: Limnophila This species is distributed as an aquarium plant but it has not yet been reported out of cultivation in BC.

{Limnobium Laevigatum} (Humb. and Bonpl.) Heine: Amazonian frogbit This species is distributed as an aquarium plant but it has not yet been reported out of cultivation in BC, [*Limnobium stoloniferum*].

{Limnobium Spongia} (Bosc) Steud: American frogbit This 'weedy' species is widely distributed as an aquarium plant but it has not yet been reported out of cultivation in BC.

Limosella Aquatica L: Water mudwort, Limosella, Mudwort Wet sites and shallow water in the lowland, steppe and montane zones; infrequent throughout BC south of 57 degrees N, absent from the Queen Charlotte Islands; circumpolar.

Lobelia Dortmanna L: Water lobelia Shallow water with a sandy, gravelly or otherwise firm bottom in lakes and ponds in the lowland zone; locally common in south-western BC.

Ludwigia Palustris (L.) Ell: Water purslane, American seedbox, Creeping primrose, False loosestrife, Low seedbox, Marsh ludwigia, Marsh purslane, Marsh seedbox, Phthisis weed Wet sites, shallow water of lakes and ponds in the lowland zone; rare on southern Vancouver Island and the lower Fraser Valley.

Lysimachia thyrsiflora L: Tufted loosestrife, Water loosestrife Shallow water of marshes, lakes and ponds in the lowland and montane zones; frequent in BC.

Marsilea Vestita Hook. and Grev: Hairy water clover, Clover fern, Hairy pepperwort, Water shamrock, Western water clover Inundated lake margins (exposed in the summer) in the steppe and montane zones; rare in south-central BC, known from around Vernon.

Megalodonta Beckii (Torr. ex Spreng.) Greene: Water marigold, Bur marigold Lake shores and shallow open water; rare known from south-eastern BC and Vancouver Island. [*Bidens beckii* (Torr. ex Spreng.)].

Menyanthes Trifoliata L: Buckbean, Bog bean, Marsh trefoil Bogs, shallow water of ponds and lakes and lake inlet and outlet fans in the lowland, montane and steppe zones; common throughout BC; circumpolar.

Montia Fontana L: Water chickweed, Blinks Wet meadows or shallow water in the lowland and montane zones; rare in north-western, south-western and south-central BC; circumpolar. *Myriophyllum* The species of this genus are commonly known as Frills, Milfoils, Parrot feathers, Quills, and Water milfoils.

{Myriophyllum Aquaticum (Vell.) Verd.}: Brazilian water milfoil, Golden myriophyllum, Parrot feather, Water feather Shallow water of garden ponds and drainage ditches in the lowland zone; infrequent on southern Vancouver Island and the lower Fraser Valley; introduced from South America.

Myriophyllum aquaticum is established in several lakes and rivers in Washington and Oregon, 2 lakes are in northwest Washington. [*Myriophyllum brasiliense* Cambess. in Hill *et al.*]

Myriophyllum Farwellii Morong: Farwell's water milfoil Lakes and sloughs in the lowland and lower montane zones; infrequent, scattered throughout southern BC.

{Myriophyllum Heterophyllum Michx.}: Various leaved water milfoil, Broad leaf water milfoil, Foxtail, Mare's tail, Mermaid weed Lakes and ponds in the lowland zone; rare, found in Beaver Lake in Stanley Park (now extirpated) and several ponds in Queen Elizabeth Park in Vancouver; introduced from eastern North America.

Myriophyllum Hippuroides Nutt: Western water milfoil, Brown myriophyllum, Red water milfoil, Variable water milfoil Lakes, ponds, sloughs and ditches in the lowland zone; infrequent, known from the lower Fraser Valley.

Myriophyllum Pinnatum (Walt.) B. S. P: Green parrot's feather, Cut leaf water milfoil, Eastern water milfoil, Variable water milfoil Lakes, ponds, sloughs and ditches in the lowland zone; infrequent, known from the lower Fraser Valley. [*Myriophyllum scabratum* Michx.]

Myriophyllum Quitense H. B. K: Waterwort water milfoil, Andean water milfoil, Coarse water milfoil Lakes and rivers in the lowland zone; rare on Vancouver Island.

Myriophyllum Sibiricum Kom: Siberian water milfoil, American water milfoil, Northern water milfoil, Spiked water milfoil, Western spiked water milfoil Lakes, ponds, rivers and other water bodies in the lowland and montane zones; frequent throughout BC; circumpolar. [*Myriophyllum exalbescens* Fern.]

Myriophyllum Spicatum L: Eurasian water milfoil, Fox tail, Spiked water milfoil Lakes, ponds, sloughs, irrigation ditches and other water bodies in the lowland, steppe and montane zones; infrequent, scattered throughout southern BC; introduced from Eurasia.

Myriophyllum Ussuriense (Regel) Maxim: Ussurian water milfoil Marginal habitats of lakes and rivers subject to seasonal water level fluctuations, and exposure in late summer, in the lowland and lower montane zones; rare and scattered on Vancouver Island, the lower Fraser Valley and south-central BC; amphiberingian.

Myriophyllum Verticillatum L: Verticillate water milfoil, Green milfoil, Whorled water milfoil Lakes and sloughs of the lowland to montane zones; frequent throughout BC; circumpolar.

Najas Flexilis (Willd.) R. and S: Wavy water nymph, Common naiad, Northern naiad, Slender naiad, Slender water nymph Ponds and marshes in the lowland, steppe and montane zones; common in southern BC; circumpolar.

Nasturtium Officinale R. Br. in W. Ait: Common water cress, True water cress Streams, ditches, swamp margins and shallow ponds in the lowland and montane zones; frequent in southern BC, rare northward; introduction from Europe. [*Rorippa nasturtium-aquaticum* (L.) Hayek]

Nitella**:** The species of this genus are commonly known as Muskgrasses, Nitellas, Sandgrasses, Skunkgrasses, and Stoneworts.

Nitella Acuminata A. Br: Very little data is available on the abundance and distribution of charophytes in BC.

Nitella Clavata Kutz: ery little data is available on the abundance and distribution of charophytes in BC.

Nitella Flexilis (L.) Ag: Very little data is available on the abundance and distribution of charophytes in BC.

Nitella Furcata (Roxb.) Ag: Very little data is available on the abundance and distribution of charophytes in BC.

Nitella Gracilis (Sm.) Ag: Very little data is available on the abundance and distribution of charophytes in BC.

Nitella Tenuissima (Desv.) Kutz: Very little data is available on the abundance and distribution of charophytes in BC.

Nuphar: The species of this genus are commonly known as Yellow lilies, Cow Lilies, Pond lilies, Water lilies, Spatterdocks and Water collards.

Nuphar Polysepalum Engelm: Rocky mountain cow lily, Indian pond lily, Western pond or water lily Lakes, ponds and slow-moving streams in the lowland, steppe and montane zones; common in BC south of 55 degrees N and west of 120 degrees W, less common northward and eastward.

Nuphar Variegatum Engelm: Bullhead pond lily, Common yellow pond lily, Variegated pond lily Lakes, ponds and slow-moving streams in the steppe and montane zones; frequent in eastern BC.

Nymphaea: The species of this genus are commonly known as Fragrant water lilies, Pond lilies, Water nymphs, Water lilies and White water lilies

Nymphaea Alba L.: European white water lily Ponds and lakes in the lowland and montane zones; rare on south-eastern Vancouver Island and at Clearwater; introduced from Europe.

Nymphaea Leibergii Morong: Pygmy water lily, Dwarf water lily, Little white water lily, Northern water lily, Small white water lily Lakes, ponds and slow-moving streams in the montane zone; rare and scattered in eastern BC.

Nymphaea Mexicana Zuccarini: Yellow water lily, Banana water lily, Yellow lotus Ponds and lakes in the lowland zone; rare on south-eastern Vancouver Island; introduced from Mexico.

Nymphaea Odorata Ait: Fragrant water lily, Alligator bonnet, American water lily, Cow cabbage, Large white water lily, Sweet scented water lily, Toad lily, Water cabbage Lakes, ponds and slow-moving streams in the lowland, steppe and lower montane zones; rare and scattered in south-western and south-central BC; introduced from Europe.

Nymphaea Tetragona Georgi: Pygmy water lily, Dwarf water lily, Little white water lily, Northern water lily, Small white water lily Lakes, ponds and slow-moving streams in the lowland and montane zones; rare and scattered from the central coast and interior northward; circumpolar.

Nymphoides Aquatica (Gmelin) O. Kuntze: Banana plant, Fairy water lily This species is found in garden ponds but has not yet been found naturalized.

{Nymphoides Cordatum} (Ell.) Fernald: Floating heart This species is found in garden ponds but has not yet been found naturalized.

Nymphoides Peltata (Gmelin) O. Kuntze: Water fringe, Floating heart This species is found in garden ponds but has not yet been found naturalized in BC, it is established in Washington State near Spokane.

Phyllospadix: The species of this genus are commonly known as Sea grasses or Surf grasses.

Phyllospadix Scouleri Hooker: Scouler's surf grass Exposed, rocky, intertidal to subtidal shores in the lowland zone; common along the coast.

Phyllospadix Serrulatus Rupr. ex Aschers: Toothed surf grass Sheltered intertidal and subtidal shores in the lowland zone; frequent along the coast.

Phyllospadix Torreyi S. Wats: Torrey's surf grass Exposed, rocky, intertidal to subtidal shores in the lowland zone; infrequent in coastal BC.

{Pilularia Americana R. Br.}: American pillwort Shallow water of ponds in the lowland zone; rare in south-western BC; introduced from south-western North America, known from a pond in the UBC botanical gardens.

{Pistia Stratiotes L.}: Water lettuce, Shell flower, Water cabbage, Water bonnet This is a free-floating tropical species introduced in the aquarium and garden pool trade but it has not yet been found out of cultivation It is not expected to overwinter outdoors in BC. (None of the authors specimens have ever overwintered successfully. There is a long list of endemic common names, mostly in southeast Asian languages and tropical American tongues).

Polygonum: The species in this genus are commonly known as Door weeds, Knot weeds, Smart weeds and Tear thumbs.

Polygonum Amphibium L: Water smartweed, Amphibious bistort, Water smartweed, Floating knotweed, Marsh smart weed, Swamp persicaria, Swamp smart weed, Water heartsease, Water knot weed, Water persicaria Shorelines, ditches and shallow water of lakes and ponds in the lowland, steppe and montane zones; common throughout BC except for the Queen Charlotte Islands and adjacent coast.

Polygonum Hyropiper L: Marshpepper smartweed, Annual smartweed, Common smartweed, Water smartweed, Marsh pepper, Water pepper

Moist ditches, shallow marginal sites and disturbed places in the lowland, steppe and montane zones; infrequent in south-western and south-central BC; introduced from Eurasia.

Polygonum Hydropiperoides Michx: Swamp smartweed, Marshpepper smartweed, Mild water pepper, Wild water pepper Wet swampy sites, shorelines and shallow marginal water of lakes and rivers in the lowland, steppe and montane zones; rare in southern BC.

Polygonum Lapathifolium L: Willow weed, Nodding smartweed. Pale smartweed Wet swampy sites, wet meadows, shorelines and shallow marginal water of lakes and rivers in the lowland, steppe and montane zones; common in south-western BC, rare elsewhere in southern and north-eastern BC.

{Pontederia Cordata Lour.}: Pickerel weed, Pikeweed, Wampee This introduced subtropical species is often found in garden pools and is known from Glen Lake in Victoria where it has overwintered successfully for over a decade, but not spread from its original site. (The author has kept a population growing, outdoors, in a tub of water for many years; it suffered severely during the cold winter of 1995 but survived and recovered completely). *Potamogeton* The species of this genus are commonly known as Fishweeds, Pondweeds and River weeds.

Potamogeton Alpinus Balbis: Northern pondweed, Alpine pondweed, Red pondweed Nutrient poor lakes and sloughs in the lowland, steppe and montane zones; frequent throughout BC; circumpolar.

Potamogeton Amplifolius Tucker: Large leaved pondweed, Bass weed, Big pondweed, Broad pondweed, Large leaf pondweed, Muskie weed Lakes and ponds in the lowland and montane zones; frequent in south-western BC, infrequent east of the Coast-Cascade Mountains south of 55 degrees N.

Potamogeton Berchtoldii Fieb. in Bercht: Berchtold's pondweed, Small pondweed Lakes, ponds, sloughs and ditches in the lowland, steppe and montane zones; common throughout BC; circumpolar.

Potamogeton Crispus L: Curled pondweed, Crimped pondweed, Crisp pondweed, Curled pondweed, Curly leaf pondweed, Curly cabbage or muckweed Nutrient rich lakes, ponds and sloughs in the lowland and steppe zones; common in BC south of 50 degrees N; introduced from Eurasia.

Potamogeton Epihydrus Raf: Ribbon leaved pondweed, Leafy pondweed, Nuttall's pondweed Peaty lakes, ditches and ponds from the lowland and steppe to subalpine zones; common in coastal BC, infrequent east of the Coast-Cascade Mountains.

Potamogeton Filiformis Pers: Slender leaved pondweed Calcium rich lakes in the lowland, steppe and montane zones; frequent east of the Coast-Cascade Mountains, rare along the coast; circumpolar.

Potamogeton Foliosus Raf: Closed leaved pondweed, Leafy pondweed, Narrow leaf pondweed Lakes, ponds and ditches in the lowland, steppe and montane zones; frequent in BC south of 55 degrees N, rare in northern BC.

Potamogeton Friesii Rupr: Flat stalked pondweed, Fries' pondweed Lakes, ponds and ditches in the lowland, steppe and montane zones; frequent in BC east of the Coast-Cascade Mountains, infrequent in coastal BC; circumpolar.

Potamogeton Gramineus L: Grass leaved pondweed, Variable leaf pondweed Lakes, lake margins, ponds in peat bogs, ditches and slow flowing streams and rivers in all but the alpine zone; frequent throughout BC; circumpolar. A very plastic and variable species often found growing as a rosette on damp mud when water levels recede.

Potamogeton Illinoensis Morong: Illinois pondweed Lakes and slowly flowing streams in the steppe zone; infrequent in southern BC east of the Coast-Cascade Mountains.

Potamogeton Natans L: Floating leaved pondweed, Bass weed, Broad leaf pondweed, Common floating pondweed, Floating brown leaf, Muskie weed Lakes, streams and ponds from the lowland and steppe to subalpine zones; common throughout BC; cosmopolitan.

Potamogeton Nodosus Poir: Long leaved pondweed, American pondweed, Knotty pondweed, Loddon pondweed, River pondweed, River weed Lakes and sloughs of the lowland zone; infrequent in the lower Fraser Valley, rare elsewhere; cosmopolitan.

Potamogeton Oakesianus Robbins: Oakes' pondweed Lakes in the lowland and steppe zones; rare in southern BC, found in Steelhead and Mara Lakes.

Potamogeton Obtusifolius Mertens and Koch: Blunt leaved pondweed, Grassy pondweed Lakes and sloughs in the lowland, steppe and montane zones; infrequent throughout BC; circumpolar.

Potamogeton Pectinatus L: Sago pondweed, Big pondweed, Large sheath pondweed, Bushy pondweed, Fennel leaved pondweed, Fine leaf pondweed, Giant pondweed, Narrow leaf pondweed, Sheathed pondweed, Slender pondweed, Thread leaf pondweed Lakes, ponds and estuaries in the lowland, steppe and montane zones; frequent in BC east of the Coast-Cascade Mountains, infrequent on the coast; cosmopolitan.

Potamogeton Perfoliatus L: Perfoliate pondweed, Bass weed, Clasping leaf pondweed, Muskie weed, Perfoliate pondweed, Redhead grass Lakes in the montane zone; rare in northern BC, found in Swan Lake in the Cassiar Range; circumpolar.

Potamogeton Praelongus Wulf: Long stalked pondweed, Muskie weed, White stem pond weed Lakes in the lowland, steppe and montane zones; frequent throughout BC; circumpolar.

Potamogeton Pusillus L: Small pondweed, Baby pondweed, Narrow leaf pondweed, Slender leaf pondweed, Small pondweed Lakes, ponds, sloughs and ditches in the lowland, steppe and montane zones; frequent in BC; circumpolar.

Potamogeton Richardsonii (A. Benn.) Rydb: Richardson's pondweed, Clasping leaf pondweed Lakes, ponds and sloughs in the lowland, steppe and montane zones; common throughout BC; amphiberingian.

Potamogeton Robbinsii Oakes: Robbin's pondweed, Fern pondweed, Robbin's pondweed Lakes in the lowland, steppe and montane zones; frequent in coastal BC, infrequent east of the Coast-Cascade Mountains.

Potamogeton Strictifolius Bennett: Stiff leaved pondweed, Narrow leaf pondweed Lakes in the lowland and steppe zones; rare in south-central and south eastern BC, found in Kawkawa, Mara and Windermere Lakes.

Potamogeton Vaginatus Turcz: Sheathing pondweed Lakes in the montane zone; infrequent east of the Coast-Cascade Mountains; circumpolar.

Potamogeton Zosteriformis Fern: Eel grass pondweed, Flatstem pond weed Lakes in the lowland, steppe and montane zones; frequent in BC east of the Coast-Cascade Mountains, infrequent in coastal BC.

Ranunculus: The species in this genus are commonly known as Buttercups and Crowfoots.

Ranunculus Aquatilis L: White water buttercup, Water crowfoot, Aquatic buttercup Lakes, ponds and slow-moving streams in the lowland, steppe, montane and subalpine zones; frequent throughout BC.

Ranunculus Circinatus Sibth: Stiff leaved water buttercup Ponds and slow-moving streams in the lowland, steppe and montane zones; common in southern BC east of the Coast-Cascade Mountains, less frequent northward.

Ranunculus Cymbalaria Pursh: Shore buttercup, Desert buttercup, Seashore buttercup Moist saline or alkaline shorelines, marshes, wet meadows and shallow marginal water of lakes and ponds in the lowland, steppe and montane zones; common in southern BC, less frequent northward; circumpolar.

Ranunculus Flabellaris Raf: Yellow water buttercup, Yellow crowfoot, Pursh's buttercup Ponds and shallow marginal water of lakes and ponds in the steppe and montane zones; rare and scattered in BC east of the Coast-Cascade Mountains.

Ranunculus Flammula L: Lesser spearwort, Creeping buttercup, Creeping spearwort Wet seepage sites along the shores of lakes, wet meadows and shallow marginal water of lakes and streams in the lowland, steppe and montane zones; common throughout BC; circumpolar.

Ranunculus Gmelinii DC: Small yellow water buttercup Ponds, shallow streams and wet sites in the lowland, steppe and montane zones; common throughout BC except rare along the coast; amphiberingian.

Ranunculus Hyperboreus Rottb: Arctic buttercup, Far northern buttercup, Floating buttercup Ponds and shorelines in the montane zone; frequent in northern BC, rare southward to 53 degrees N; circumpolar.

Ranunculus Lobbii (Hiern) A. Gray: Lobb's water buttercup Vernal pools and wet sites in the lowland zone; rare on south-eastern Vancouver Island, not collected for 50 years.

Ranunculus Sceleratus L: Cursed buttercup, Cursed crowfoot, Celery leaved buttercup, Celery leaved crowfoot Ponds and wet sites in the lowland, steppe and montane zones; common in southern BC east of the Coast-Cascade Mountains, infrequent elsewhere.

Riccia Fluitans L: Crystalwort, Dissected liverwort, Floating liverwort, Floating crystalwort This native aquatic liverwort is rarely collected or reported in aquatic habitats, but is known from a number of scattered sites in BC.

Ricciocarpus Natans (L.) Corda: Purple fringed riccia, Ricciocarpus This native liverwort is rarely collected or reported in aquatic habitats, but is known from a number of scattered sites in BC. (The world distribution is cosmopolitan; the author has seen it growing with *Victoria amazonica* in the Peruvian Amazon basin and noted it in many eutrophic sites throughout BC).

Ruppia Maritima L: Ditch grass, Sea tassel, Widgeon grass Tidal marshes, ponds, lakes and ditches in the lowland, steppe and montane zones; common along the coast and in south-central BC, rare in south-eastern BC; circumpolar. A very variable species or species-complex

Sagittaria: The species in this genus are commonly known as Arrow heads, Duck potatoes, Swamp potatoes and Wapato.

Sagittaria Cuneata Sheldon: Arum leaved arrowhead, Northern arrowhead, Western wapato Wet ditches, ponds and lakes in the lowland, steppe and montane zones; common in BC south of 55 degrees N and east of the Coast-Cascade Mountains, less frequent westward and northward.

Sagittaria Latifolia Willd: Common arrowhead, Broadleaf, Coastal and Slender arrowhead Wet ditches, ponds, lake shores and marshes in the lowland, steppe and montane zones; uncommon in BC south of 56 degrees N, absent from the Queen Charlotte Islands, Northern Vancouver Island and the adjacent coast.

Salvinia: The species in this genus are commonly known as Floating ferns, Floating mosses, Salvinia, Water ferns and Water spangles. Most specific common names are in southeast Asian dialects where the plants are rampant. Introduced tropical free-floating ferns present in the aquarium and water garden trade, occasionally dumped in local lakes but not yet known, or expected, to overwinter in BC. (Although specimens the author has grown in outdoor pools have never overwintered successfully, spores are produced and should be able to survive the winter in suitable habitats.

One species, {*Salvinia natans*}, is known from Europe and is capable of growth in temperate conditions, these plants could become naturalized in BC. {*Salvinia auriculata*} [*Salvinia molesta*] is a tropical species complex, none of which are likely to become established).

Scheuchzeria palustris L: Scheuchzeria Shallow marginal water of lakes and ponds and bogs in the lowland and montane zones; frequent in south-western, central and eastern BC, also along the coast.

Scirpus Lacustris L: American bulrush, Big bulrush, Giant bulrush, Greater bulrush, Porcupine quill bulrush, Slender bulrush, Softstem bulrush, True bulrush, Zebra rush, Zebra bulrush, Tule Shallow water of marshes, lakes and streams in the lowland, steppe and montane zones; common throughout BC. (A common species-complex throughout BC which includes *Scirpus validus* Vahl and *Scirpus acutus* Muhl. in Bigel.).

Scirpus Subterminalis Torr: Water clubrush, Aquatic sedge, Swaying rush Shallow ponds and streams in the lowland zone; infrequent in coastal and south-central BC *Sparganium* The species in this genus are commonly known as Bur reeds.

Sparganium Angustifolium Mich: Narrow leaved bur reed, Floating leaf bur reed, Green fruited bur reed, Slender bur reed, Unbranched bur reed, Western bur reed Ponds, ditches and lake shores in all but the alpine zone; common throughout BC; circumpolar. A highly variable species or species-complex which sometimes includes *Sparganium emersum* as well.

Sparganium Emersum Rehm: Emersed bur reed Streams, ditches, ponds and lake shores in the lowland, steppe and montane zones; common in coastal BC, less frequent elsewhere; circumpolar.

Sparganium Eurycarpum Engelm. in A. Gray: Broad fruited bur reed, Big, .Common bur reed, Eastern bur reed, Giant bur reed, Large bur reed, Three square bur reed Wet meadows, shallow ponds and lake shores in the lowland, steppe and montane zones; infrequent in southern and eastern BC.

Sparganium Fluctuans (Morong) B. L. Robins: Water bur reed, Broad ribbon leaf, Floating leaf bur reed, Shining bur reed Ponds, lake shores and slow-moving streams in the lowland and montane zones; rare in south-western and central BC; circumpolar.

Sparganium Glomeratum Laest. ex Beurl: Glomerate bur reed Ponds and lake shores in the lowland and montane zones; rare in west-central BC, known only from Buck Channel, Queen Charlotte Islands and the Smithers area.

Sparganium Hyperboreum Laest. ex Beurl: Northern bur reed Ponds and lake shores in the lowland and montane zones; infrequent throughout BC; circumpolar.

Sparganium Natans L: Small bur reed Ponds, lake shores and slow-moving streams in the lowland and montane zones; common throughout all but north-western BC; circumpolar. [*Sparganium minimum*].

Spirodela Polyrhiza (L.) Schleid: Great duckweed, Big, Greater duckweed, Larger duckweed Ponds, lakes and slow-moving streams in the lowland, steppe and montane zones; common in south-western and south-central BC, infrequent elsewhere, absent on northern Vancouver Island, the Queen Charlotte Islands and north-western BC; circumpolar.

Subularia Aquatica L: Water Awlwort Streams, shorelines, tidally inundated shores, shallow ponds and lakes; infrequent in southern BC.

Tolypella Intricata (Trent.) Leonh: Very little data is available on the abundance and distribution of charophytes in BC.

Trapa Natans L.: Water chestnut, Caltrops, Water nut This species has not yet been found in BC but if introduced it should be able to overwinter and could become a pest.

Utricularia: The species in this genus are commonly known as Bladderworts.

Utricularia Gibba L: Humped bladderwort, Cone spur bladderwort, Eastern bladderwort, Small bladderwort, Yellow bladderwort Lake bottoms and margins, muddy disturbed sites in the lowland zone; rare on southern Vancouver Island and in the Vancouver area.

Utricularia Intermedia Hayne: Flat leaved bladderwort, Mountain bladderwort Oligotrophic and dystrophic lakes and marshes in the lowland, steppe and montane zones; frequent south of 55 degrees N, rare northward; circumpolar.

Utricularia Minor L: Lesser bladderwort, Northern bladderwort, Small bladderwort Oligotrophic and dystrophic lakes and peat bog pools in the lowland and montane zones; common on Vancouver Island and the Queen Charlotte Islands, less frequent inland; circumpolar.

Utricularia Vulgaris L: Greater bladderwort, Common bladderwort Lakes and ponds in the lowland, steppe and montane zones; common south of 55 degrees N in BC, less frequent northwards; circumpolar. *Vallisneria* The species in this genus are commonly known as Eel grasses, Ribbon grasses, Tape grasses and Water celery's.

Vallisneria Americana Michx: American tapegrass Lakes, ponds and streams in the lowland and montane zones; rare in south-western and south-central BC; introduced from eastern North America.

Veronica: The species in this genus are commonly known as Brooklimes and Speedwells.

Veronica Anagallis-aquatica L: Blue water speedwell, Brook pimpernel Wet ditches and, shallow marginal water of streams and lakes in the lowland zone; rare in southern and central BC; introduced from Europe.

Veronica Beccabunga L: American brooklime, American speedwell Wet ditches, stream and lake edges and shallow marginal water in the lowland, steppe and montane zones; common in southern BC, infrequent northward. [*Veronica americana* Schwein.]

Veronica Catenata Pennel: Pink water speedwell, Tufted water speedwell Wet ditches, slow streams, stream and lake edges and shallow marginal water in the lowland and montane zones; rare in southern BC; circumpolar.

Veronica Scutellata L: Marsh speedwell Swamps, stream and lake edges and shallow marginal water in the lowland, steppe and montane zones; frequent throughout BC.

Wolffia: The species in this genus are commonly known as Water meals.

Wolffia Borealis (Engelm. ex Hegelmaier) Landolt and Wildi: Northern water meal, Dotted water meal, Papillary water meal, Spotted water meal, Southern water meal Ponds, lakes and slow-moving streams in the lowland and montane zones; rare in the lower Fraser Valley and south-eastern BC (Creston). This species is probably more common than its collection reports indicate.

Wolffia Columbiana Karsten: Columbian water meal Ponds and lakes in the lowland zone; rare in south-western BC, known only from Beaver, Blenkinsop and Swan Lakes. This species is probably more common than its collection reports indicate.

Wolffiella Floridana (J. D. Smith) Thompson: Florida water meal This species has been tentatively identified by the author from a lake near Smithers but no specimens exist for confirmation.

Zannichellia Palustris L: Horned pondweed, Common poolmat, Grass wrack Shallow, often calcareous or brackish lake shores, slow-moving streams and tidal marshes in the lowland, montane and steppe zones; frequent in mainland BC south of 54 degrees N, less frequent northward and on Vancouver Island; cosmopolitan.

Zostera: The species in this genus are commonly known as Eel grasses.

Zostera Japonica Ascher and Grabn: Japanese eelgrass Tidal mud-flats along the coast in the lowland zone; rare in south-western BC, known only from Boundary Bay and Tsawwassen; introduced from Asia.

Zostera Marina L: Common eelgrass, Wrack Sheltered coastal waters in the lowland zone; common in coastal BC; cosmopolitan.

Appendix: Collecting and Preserving Aquatic Plants

- Introduction
- General Specimen Handling Procedures
- Collecting
- Mounting and Labelling
- Emergent and Wetland Plants

- Small Floating Plants
- Submersed Flaccid and Dissected-leaf Plants
- Plants Requiring Special Handling
- Biomass Studies
- Tissue Analyses.

The collection of aquatic plants is often done as part of ecological or impact studies, both as a record of conditions at a given time for comparison with prior or later conditions, and as a necessity when the collectors are not able to identify the specimens in the field and need to send a specimen to an expert. These specimens are a valuable scientific records and their collection and subsequent handling should be done with care so that the time and expense that has gone into their collection is not wasted.

Herbaria exist for the purpose of long term care and storage of the specimens and uniform standards have been set up for the mounting, care and storage of plant specimens. These standards should be followed so that the plants will become valuable scientific specimens for more than the specific purpose for which they were collected. In addition, their storage in a long term facility like an herbarium will make them accessible to subsequent researchers for other uses. Herbaria provide a permanent record of what was found and if identifications need to be rechecked the specimens are available. Even when the major thrust of the work is for biomass studies or tissue analyses, representative specimens must still be collected and saved as vouchers of what species were analyzed or studied.

There is a good manual published by the BC Ministry of Forests (Aikens, 1996) which is recommended as a reference for collecting, preserving, processing and storing botanical specimens. Most of that information is not repeated here. This appendix is concerned only with the idiosyncrasies of certain aquatic plants which require special handling in order to make good herbarium specimens.

General Specimen Handling Procedures

Plants are mounted on white card stock, about 11 by 16 inches and stored in Herbarium Cabinets made for this purpose. These cabinets have insect-proof tightly-sealed doors, and shelves which accommodate these standard-sized mounting boards. This uniformity facilitates exchange of specimens between institutions and a common design of storage cabinets and facilities. One should assume that a specimen is both permanent and valuable; treat it as such when you collect it, when you are processing and storing it and when you are handling it at a later date. Many specimens in herbaria around the world are hundreds of years old and are of great value in documenting changes in the habitats of areas, and in the study of the ongoing process of plant evolution.

For terrestrial plants, where the plants are rigid and can be mounted dry, most good quality white card stock is acceptable. However most submersed plants do not have structural supporting tissues, they rely on the water for support, and they must be floated onto the card stock. Therefore a paper which remains dimensionally stable after wetting and subsequent drying is required. Most ordinary papers, after being soaked, will wrinkle upon subsequent drying, even if they are kept pressed during the drying process. Ask your paper supplier for a card stock which does not react this way; they are available but their trade names change from time-to-time and supplier-to-supplier.

Collecting

There are a number of aquatic plant collection methods which have been used. The one to use in a given situation depends upon the specific situation and the reason for the collection.

Picking the whole plant by hand is the best method for getting an entire plant in undamaged condition, particularly for preservation as a herbarium specimen. In shallow water this is readily done by wading if the bottom is firm, or by leaning over the side of a canoe where the bottom is too soft for wading. In deeper water snorkeling is the best method. The closer and clearer view of the bottom will often allow one to see small species that were not visible from the surface. In still deeper water SCUBA, or surface-supplied-air diving may be necessary. If there is any wave action, surface ripple or glare, it is difficult to see the bottom from the air; this problem is eliminated when one is underwater.

Rakes or cultivators with long handles can be used to uproot and bring up plants when working from a boat. Cultivators with four long, closely-set prongs are better for uprooting plants but, for small species like Isoetes, it may be difficult to bring the plants to the surface. An Ekman dredge will also bring up small species from deep water if the sediment surface is not too hard. In deep water where visibility is poor one can get a random sample of some of the plants which are present by dragging an anchor from a boat. The plants are not in very good shape and the sample is not necessarily representative of all the species present. Usually all that is collected are fragments of plants without roots or rhizomes and small plants like *Isoetes* will be missed. For sample completeness and specimen quality none of these methods are as good as picking by hand.

Generally the whole plant should be collected. Some groups can not be identified to species without mature fruits or flowers; others need rhizomes, leaf axils or tips. Since submersed aquatic plants do not have to guard against loss of water from their tissues, they do not have waxy or water-repellent cuticles like emergent plants. Do not leave them exposed, even briefly, since they will wither very quickly and become useless as specimens. Keep them in a bag or bucket of water at all times until you are ready to press them. Emergent plants should not be submersed but kept in a bag with a little water in the bottom to maintain a high humidity. It is best to keep each species in its own bag and all the bags from one lake or site together in one large bag.

Some very small plants like the duckweeds do not make very satisfactory pressed and dried specimens, nor is it convenient to collect them into a bag. Small, 20 mL, screw cap vials make good collecting and preservation containers for these plants. Put a little water into the vial to keep the plants moist. Later fill the vial with a solution of 5% formalin, 25% water and 70% ethanol as a permanent preservative. This liquid may need to be replaced after about a month since it will extract chlorophyll and pigments from the plants and become quite dark or opaque. Isopropanol can be used instead of ethanol but ethanol is preferred.

Bibliography

Ailstock, M. S.: *Utilization and Treatment of Thermal Discharge by Establishment of a Wetlands Plant Nursery,* Lewis Publishers, Chelsea, Michigan, 2009.

Aston,H.I.: *Aquatic Plants of Australia*, Melbourne University Press, Australia, 2003

Ball, J., R. D. Bauer, K. Vermeer, and M. J. Rabenberg: *Northwest Riverine and Pacific Coast*, North America. Texas Tech. Univ. Press, Lubbock, 2019.

Boyer , J.S.: *Measuring the Water Status of Plants and Soils,* Academic Press, N.Y., 2015.

Chadde,S.W.: *A Great Lakes wetland flora: A complete illustrated guide to the aquatic and wetland plants of the Upper Midwest*, Pocket Flora Press, Calumet, Michigan, 2018.

Chadha K. L. and Pareek O. P.: *Advances in Horticulture: Fruit Crops,* New Delhi, Malhotra Publishing House, 2013.

Cook, C.D.K.: *Water Plants of the World*, Dr W Junk Publishers, The Hague, 2004.

Cronquist, A.: *The Evolution and Classification of Flowering Plants*, New York Botanical Garden, Bronx, New York., 2008

David White: *The Physiology and Biochemistry of Prokaryotes*, Oxford University Press, Delhi, 2020.

Doijode S. D.: *Seed Germination in Fruits*, New Delhi, Malhotra Publishers, 2013.

Ferentinos L.: *Proceeding of the Sustainable Taro Culture for the Pacific Conference*, Honolulu, HITAHR, 2013.

Godden G.: *Growing Citrus Trees*, Australia, Lothian Publishing Company Pvt. Ltd., 2008.

Godfrey,R.K. and Wooten,J.W.: *Aquatic and Wetland Plants of Southeastern United States*, University of Georgia Press, Athens, 2009.

Herminie Broedel Kitchen: *Soils and Crops : Diagnostic Techniques*, Satish Serial Publishing, Allahabad, 2018.

Hutchinson, G. E.: *Limnological Botany*, John Wiley, New York, 2005.

Jones, R. M.: *Plant Resources of South-East Asia,* Wageningen, Pudoc Scientific Publishers, 2012.

Keddy, P.A.: *Principles and Conservation,* Cambridge University Press, Cambridge, UK, 2023.

Keddy, P.A.: *Wetland Ecology: Principles and Conservation,* Cambridge University Press, Cambridge, UK, 2020.

Lorenzen, S.: *The Phylogenetic Systematics of Freeliving Nematodes*, London, The Ray Society, 2014.

Madulid Domingo A.: *A Pictorial Cyclopedia of Philippine Ornamental Plants*, Philippines, Makati Metro Manila, 2015.

Mason,H.L.: *A flora of the Marshes of California*, University of California Press, Berkeley, 1997.

Nobel, P. S.: *Physicochemical and Environmental Plant Physiology*, Academic Press, San Diego, 2009.

Pemberton, R. W.: *Predictable Risk to Native Plants in Weed Biological Control,* Oecologia, 2017.

Pilgrim R.: *Post Harvest Handling of Minor Exotics*, St. George's, Grenada, 2016.

Pramila Tripathi: *Botanical Pesticides in the Management of Post-harvest Fruit Diseases*, Daya, Delhi, 2015.

Rehm Sigmund: *Multilingual Dictionary of Agronomic Plants*, Boston, Kluwer Academic Publishers, 2014.

Rich, T.C.G. and Jermy, A.C.: *Plant Crib 1998*, Botanical Society of the British Isles, London, 2008.

Rutherford Lyn.: *A Gourmet's Book of Mushrooms & Truffles*, Sydney, Golden Press Pvt. Ltd., 2011.

Sanchez, P. A.: *Properties and Management of Soils in the Tropics*, John Wiley and Sons, New York, 2006.

Sculthorpe, C. D.: *The Biology of Aquatic Vascular Plants*, Cambridge University Press, London, 2007.

Tomlinson, P. B.: *The Botany of Mangroves*, Cambridge University Press, Cambridge, UK, 2016.

Westlake, D.F., Kvt, J. and Szczepañski, A.: *The Production Ecology of Wetlands*, Cambridge University Press, Cambridge, 2018.

Whealy K.: *The Garden Seed Inventory*, Decorah, Seed Saver Publications, 2018.

White, G.F.: *Natural Hazards: Local, National, Global*, Oxford University Press, New York, 2014.

Index